MULTICULTURAL CITIES

Toronto, New York, and Los Angeles

What defines a multicultural city? Policy? Geography? Demography? In *Multicultural Cities*, Mohammad Abdul Qadeer offers a tour of three of North America's largest multicultural metropolises – Toronto, New York, and Los Angeles – that demonstrates the essential qualities that make these cities multicultural.

Guided by the perspective that multiculturalism is cultural diversity supported by a common ground of values and institutions, Qadeer examines the social geography, economy, and everyday life of each metropolitan area. His analysis captures the important differences between cities in Canada, where multiculturalism is official government policy, and in the United States, where it is not. Written by a keen observer of North American urban life, *Multicultural Cities* is a comprehensive investigation of how some of today's leading majority-minority cities thrive.

MOHAMMAD ABDUL QADEER is a professor emeritus in the Department of Geography and Planning at Queen's University.

MOHAMMAD ABDUL QADEER

Multicultural Cities

Toronto, New York, and Los Angeles

UNIVERSITY OF TORONTO PRESS
Toronto Buffalo London

© University of Toronto Press 2016
Toronto Buffalo London
www.utppublishing.com

ISBN 978-1-4426-3013-0 (cloth)   ISBN 978-1-4426-3014-7 (paper)

Library and Archives Canada Cataloguing in Publication

Qadeer, Mohammad A., author
Multicultural cities : Toronto, New York, and Los Angeles / Mohammad
Abdul Qadeer.

Includes bibliographical references and index.
ISBN 978-1-4426-3013-0 (cloth). – ISBN 978-1-4426-3014-7 (paper)

1. Multiculturalism – Ontario – Toronto.   2. Multiculturalism – New York (State) –
New York.   3. Multiculturalism – California – Los Angeles.   I. Title.

HN110.T6Q33 2016   305.8009713/541   C2015-908249-8

This book has been published with the help of a grant from the Federation for the
Humanities and Social Sciences, through the Awards to Scholarly Publications
Program, using funds provided by the Social Sciences and Humanities Research
Council of Canada.

University of Toronto Press acknowledges the financial assistance to its publishing
program of the Canada Council for the Arts and the Ontario Arts Council, an
Ontario government agency.

Canada Council   Conseil des Arts
for the Arts   du Canada

ONTARIO ARTS COUNCIL
CONSEIL DES ARTS DE L'ONTARIO
an Ontario government agency
un organisme du gouvernement de l'Ontario

Funded by the   Financé par le
Government   gouvernement
of Canada   du Canada

To my grandchildren

Contents

Maps

Photographs

Tables

Preface

Cities have been described in kaleidoscopic terms. Look from one perspective and cities appear to be commercial, industrial, or post-industrial; examine them from another angle and they fall into categories of regional, national, and global; or they can be classified as post-modern, creative, wired, or fragmented. This proliferation of urban labels is not a failure of theoretical consistency. Rather, it is an acknowledgment of the multiplicity of aspects that constitute the reality of cities.

Multicultural city is a relatively new term. It focuses on the cultural and social aspects of city life. It refers to the pluralistic society emerging from a combination of ethno-cultural diversity and equality rights. It points to a specific form of urban culture and society that comes into being with the infusion of pluralism into a city's social, economic, and political institutions. A multicultural city rises on the foundations of an existing city and thus is embedded in its history and geography. Yet it has a distinct spirit of not just tolerating but honouring cultural differences.

This book is an inquiry into the institutions, structures, and processes of multicultural cities. It draws on the experiences of three leading and recognized multicultural cities of North America, namely, Toronto, New York, and Los Angeles. The broad question addressed in this book is how the ethno-racial diversity of a city's population is embedded in its geographic structure, social organization, economic activities, and political and symbolic institutions, all as a matter of rights. The book argues that the diversity promoted by multiculturalism operates within the imperatives of living together as a civic community in a defined space. This model recognizes the enactment of ethno-racial groups' cultural rights in their individual and community life, on the one hand,

and envisages a vigorous common ground of shared land, history, values, laws, interests, technology, and ethics that integrates diverse communities into a coherent civic society, on the other. The two sides of multiculturalism complement each other. The challenge of multiculturalism lies in keeping a moving equilibrium between diversity and the evolving common ground. Multicultural cities are the stage where this process plays out.

Cities have long drawn people of different races, ethnicities, beliefs, and often languages. Social heterogeneity is their defining characteristic. Multiculturalism takes the heterogeneity to another plane. It confers equality rights on ethno-racial minorities and extends recognition and legitimacy to cultural/religious differences. Multiculturalism does not have to be proclaimed as an official policy. It also comes into existence as a lived reality with the accommodations of diversity. That is why American cities are multicultural by practice.

A multicultural city has thriving ethnic enclaves and economies, open neighbourhood and housing, religious freedom, colourful public spaces throbbing with the variety of dresses, foods, faces, music, art, and accents as well as fairs and parades of many nationalities. In it, paths are opened for minorities' representation in politics and authority, provisions are made for culturally and ethnically sensitive services, policies are instituted to introduce inclusive educational offerings, and the public discourse is broadened to cultivate an ethics of trust and tolerance. All these characteristics are the outcomes of the recognition and (reasonable) accommodations of racial and cultural differences. These qualities are not realized without conflicts and controversies. The path of multiculturalism is not without twists and turns. The movement towards the realization of a multicultural city can be reversed as the political and economic conditions change.

Multiculturalism as an ideal counters the expectation of a uniform urban culture, and posits a federated urban culture, where the (sub) cultural differences of communities are nested in the civic culture that incorporates the shared interests and rights of citizenship. This book focuses on analysing the emergence of such urban cultures in some North American cities.

Methodologically, this book is a work of synthesis. It draws primarily on published scholarly and analytical studies to compose an account of the multicultural aspects of the three cities. For Toronto, I have done some empirical studies of my own that have given me an understanding of the city's social geography, ethnic economies, and planning policies. I have also relied on census data and other official sources of demographic and economic indicators for the three cities. Ethnographic accounts and newspaper stories have provided a human face for the observations drawn from the data-based studies. This book is not based on original data collected specifically for it. Yet the information and observations drawn from the secondary sources have been critically assessed for their reliability and validity.

The sources of information and data are from two different countries, namely, Canada and the United States. There are some variations in definitions and the contexts of the two. For example, the Canadian and US decennial censuses are taken in proximate but different years. Canadian census data are collected in the first year of every decade, while the US census is based on the information gathered in a decade's last year. Also, Canada has had a mid-decade census that the United States did not, though it has been substituted with a voluntary National Housing Survey. The US Census Bureau offers a three-year American Community Survey data set. Comparisons take into account differences of indicators and sources.

Another point worthy of notice is that this is a book about *multiculturalism* in cities. It focuses on the multicultural dimensions of the three cities' institutions and structures. It is neither a comprehensive analysis of the cities, nor is it about the settlement and integration of immigrants. Undoubtedly, immigration is the driving force of cultural and racial diversity in contemporary cities, but multiculturalism is a distinct phenomenon. It is about the incorporation of diversity into urban societies. It is driven equally by the native-born second and third generations of immigrants and racial minorities. Thus, the issues of immigrants' identity and integration overlap, but are not synonymous with multiculturalism. This distinction must be kept in mind.

A clarification about my approach to the referencing of sources is necessary. I have extensively footnoted the sources of my ideas and evidence. Yet in order to avoid overburdening the text with footnotes in a paragraph paraphrasing an idea from a single source, sometimes only the most critical sentence has been referenced.

Acknowledgments

It is a cliché to say that one needs a village to raise a child. But it is also true that many hands write books, even though they carry the author's name. The eloquent testimonials to the contributions of friends and family in acknowledgments bear the evidence of such group efforts. This book is certainly the product of many minds, though I bear the responsibility for its shortfalls. My wife, Susan Qadeer, has read, critiqued, and improved its successive drafts. Her experience in counselling students of many nationalities and origins in post-secondary institutions has been a rich source of insights, as have her editorial skills, which saved me from grammatical mistakes. Her intellectual contributions are over and above her love, support, and care, which have defined our life together. Sandeep Kumar Agrawal, University of Alberta, has been my partner in research. We have published many joint papers about the multiculturalism of Toronto. Without his support, I might not have travelled this research path and probably would not have gained the confidence to undertake this book project. More specifically, he reviewed many of the chapters and gave valuable suggestions.

Ted Richmond has shared generously his deep knowledge and long experience of involvement in the policies concerning immigrants' settlement and ethnic relations in Toronto. He has read and commented on the drafts of all the chapters. His encouragement and suggestions have made a difference in the text. I am particularly grateful to him for preventing me from lapsing into insensitivity in the use of terms.

David Gordon is a friend of many years. His comments on a few chapters helped sharpen my arguments. He took great interest in my book project and extended the support of the School of Urban and Regional Planning, Queen's University, as its director. Arun Peter Lobo of the Department of City Planning, City of New York, has generously responded to my requests for demographic data and maps of the city's ethno-racial profile.

My interest in the multiculturalism of cities was ignited by my graduate students' research about some aspects of immigrants' and native-born ethnics' role in restructuring Canadian cities. Lisa Domae, Kelly Grover, Maghfoor Chaudhry, and Neelam Bangash explored a variety of topics highlighting the processes of imprinting cultural and racial identities on the institutions of Canadian cities. In one sense, this book carries forward the themes they initiated.

I have had lifelong institutional support from my academic home, the School of Urban and Regional Planning, Queen's University, Kingston, Canada. It has continued in one form or another after my retirement. I have explored my ideas with students in my yearly short course on planning for multicultural cities at Queen's. In Toronto, two institutions served as my base. I have been associated with the Joint Centre of Excellence for Research on Immigration and Settlement (CERIS) at the University of Toronto, which gave me access to its library, seminars, meetings, and network of researchers. These resources proved to be very valuable in my study. Through the centre, 1 also gained access to the libraries of the University of Toronto, which are unsurpassed in their holdings of materials of interest to me. The writing of this book would have been more demanding without these resources. I had an affiliation with the School of Urban and Regional Planning, Ryerson University, as an adjunct for a few years, which brought me into contact with students and faculty. This opportunity allowed me to hone my ideas about multiculturalism. I owe gratitude to all these institutions.

The anonymous reviewers of my manuscript made insightful comments. I have taken those into account and am grateful for their suggestions.

Jo-Anne Tinlin has corrected and formatted my typescript. Thanks to her for a long-lasting friendship and readiness to help.

There are many others whose contributions have indirectly enriched the book. They have freely shared their thoughts and observations on topics that I happened to be struggling with at a given moment. My daughter Nadra Qadeer's passionate arguments about refugees' rights and their entitlement to fair treatment have kept me attuned to the less noticed dimensions of ethno-racial politics. Discussions with my son Ahmer Qadeer gave me a feel for the revelling in differences that defines everyday life in New York and brought to my attention new literary currents relevant to the city's life. My son Ali Qadeer's aesthetic sensibilities and his engagement with urban issues have enhanced my appreciation of the role of signs and symbols in knitting places together. He has also helped in the layout of maps in this book. My children's partners, Millie Dorregaray and Laura Fisher, and my grandchildren have enriched my life with their presence and affection. There are many others whose remarks and ideas have alerted me to some overlooked aspects of my research, for example, the anonymous readers of my articles, participants in conferences, and acquaintances who shared their stories of living in Toronto, New York, or Los Angeles. They remain unnamed, but not unacknowledged. I also am grateful to Douglas Hildebrand, acquisitions editor with the University of Toronto Press, for his support and persistence in this project.

MULTICULTURAL CITIES

Toronto, New York, and Los Angeles

Chapter One

Cultures and the City

Cultural Diversity in the City

Cities have been places where a variety of beliefs, behaviours, languages, and lifestyles have flourished. Social heterogeneity is the defining characteristic of a city. A walk down the street of an ancient city would have brought one face to face with people of different looks, dresses, and languages. Medieval cities were divided into quarters of distinct occupational castes, guilds, and ethnicities. Driven by mobility and communication, modern cities harbour a global cast of racial and cultural groups. But at this point in history, the early twenty-first century, Western cities in general and North American cities in particular have become crucibles of multiculturalism. They are increasingly becoming cities where the multiplicity of cultures thrives and diversity of ethnicities as well as lifestyles reigns, supported by civil rights.

Each of New York, Los Angeles, London, Toronto, or Sydney has been described as a multicultural city. One popular indicator of their multiculturalism is the presence of restaurants offering French, Mexican, Chinese, Indian, Thai, Japanese, and many other fares. The BBC radio's World Service holds a call-in competition for listeners to make the case for the most multicultural city in the world.[1] The official website of the City of Toronto declares it to be "one of the most multicultural cities in the world."

The diversity of multicultural cities is largely attributable to the flow of immigrants and sojourners from far and near. That people of different cultures come to live in these cities with the entitlements of citizenship and civic rights is what differentiates them from the multi-ethnic cities of the past and present. To be diverse is increasingly constituent of multicultural cities.

Multiculturalism is primarily an urban phenomenon, because here different cultures coexist in a closely defined space. The multicultural city as a type of urban settlement has entered the public discourse only recently, though in some form it has existed in all times. An exploration of the phenomenon of the multicultural city is in order at the present juncture in urban theory.

Contemporary multiculturalism is embedded in the ideologies of the modern state and citizenship. It is based on the rights of citizens to equality before the law and freedom of association, mobility, residence, religion, expression, and from discrimination. Furthermore, these rights apply both to individuals and groups. They also arise from the international movement towards enshrining in national constitutions the rights of regional and ethnic minorities. Yet these rights are not absolute and are circumscribed by national ideologies and laws. They can be curtailed, though not withdrawn, as in the current concerns about national security and terrorism in North America and Europe. Nonetheless, those rights confer on individuals and groups the entitlements of organizing their domestic and community life in accordance with their cultures and identities, and demand accommodations in the public sphere.

It is the regime of rights that distinguishes the "new" multiculturalism of the post-1960s era from the "old" multiculturalism of paternalist tolerance of people of different cultures or medieval examples of self-regulating communities under a king/emperor. Multiculturalism should be distinguished from multi-ethnicity, which can exist without equal rights and has been a long-standing feature of human societies.

Undoubtedly, the notion of a multicultural city does not go unchallenged. Even the idea of multiculturalism has its critics, who view it as a threat to national identity and cohesion and see it as the basis of segregating minorities. Cultural differences are regarded as transitory, meant to be obliterated by the assimilation of immigrants and minorities into the mainstream. These arguments are based on the notion of the unity of national culture and society, an assumption that is not borne out by either historic or contemporary experiences. They are also based on the "complicated fear of the other,"[2] the long-established residents' fear of losing identity, jobs, and privileges. I will take up these arguments in the next chapter when I examine the scope and limits of multiculturalism within a city. For the time being, it is enough to note that the concept of a multicultural city as a category of cities has its critics, inspired by ideological preferences for a unitary state and society.

The Multicultural City

This project proceeds from the assumption that the multicultural city is a distinct form of human settlement. Toronto or New York, for example, are multicultural cities because they allow associations, communities, clubs, and other institutions to be formed along cultural, religious, and linguistic lines; promote public expressions of community cultures; support neighbourhoods dominated by one or another ethnic group; sustain the diversity of their residents as a matter of civic rights; and accommodate cultural differences in the provision of educational and public services. They also subscribe to the notion of racial equality. Cities that do not promote civic rights to diversity are not multicultural even if people of different cultural origins live there, as is the case of Saudi Arabian and Gulf cities with large expatriate populations. They are multi-ethnic without being multicultural. When the multiplicity of community cultures are woven into its institutions and imprinted on its geography, a city becomes multicultural.

A multicultural city is a place where different races, cultures, and lifestyles are valued; their differences are recognized, accommodated, and interwoven into a shared citizenship and common institutions. It does not have to be proclaimed by a mayor or city council. It is a lived condition that prevails when residents' cultural, linguistic, and religious freedoms are protected by a nation's constitution, civic and human rights, local regulations, and due process. The United States is not a country that subscribes to the ideology of multiculturalism. It has been described as a "melting pot." Yet it is multicultural for all intents and purposes. Nathan Glazer has conceded that America has passed "beyond the melting pot." It has to accommodate multiculturalism, he says.[3] Nancy Foner argues that New York has evolved its unique brand of multiculturalism.[4] Los Angeles has long been branded a multicultural city. The changing demography of the country has reframed the sociology of race and ethnicity, making it a multicultural country.[5] These US cities are as multicultural in lived reality as is Toronto, which embodies the national policy of multiculturalism.[6] Now that Britain has been retreating from official multiculturalism, does it mean that London will cease to be a city of many cultures?

The "city" is not just a passive stage for the drama of multiculturalism. It has an active role in giving form and substance to multiculturalism. Its defined space, services, and institutions channelize cultural diversity and imprint the long-established structures of a city with pluralistic forms, balancing civic continuity and cultural change. It builds a new aesthetics of the city on its historical foundations. To a resident waking up after half a century of sleep, a multicultural Toronto or New York would be dazzling but not unrecognizable.

The contemporary multicultural city thrives on immigrants and sojourners, though the cultural diversity that they spawn is reinforced now by the forces of globalization. The multiplicity of its cultures comes from people of distant lands who have made it their home and from the natives over whom the cosmopolitanism of the global economic order has swept. The spirit of a multicultural city is captured in the phrase "the world in a city."

Immigrants of today are the citizens of tomorrow, turning from settlers to ethnics. They speak their mother tongues, practise their religions, and organize their households and community life according to their identities and values. Similarly, free trade, jet travel, the Internet, and satellite television are continually funnelling visitors, new ideas, and different lifestyles into cities. These sources transplant new beliefs and behaviours, giving another push towards the diversity of cultures within a city.

The literature on multicultural cities emphasizes their internal differences. Yet there is much that ties them together. The question that needs to be raised is how these cities function and cohere, accommodating differences and integrating diverse cultures. The salience of a multicultural city lies as much in its differences as in what makes it a whole and how that comes about. Furthermore, the cultural diversity of multicultural cities is continually evolving, as old differences fuse into each other and new modes of differentiations arise. This dynamic of fusion and divergence needs to be examined. These themes will be the focus of this book. They will be examined from two angles: first, how cultural differences are institutionalized; second, how these differences are bridged to form an integrated city.

The Book

The focus of this book is on identifying the structures, institutions, and processes of a typical multicultural city, particularly in North America, that is, Canada and the United States. It aims at describing and explaining those features of cities which represent the influences of multiculturalism. Specifically, it examines the three acknowledged and premier multicultural metropolises of North America, namely, Toronto (Canada), New York, and Los Angeles (USA), to observe how the ethno-racial diversity of their residents has been accommodated in their social geography, economic organization and entrepreneurial culture, organization of communities, provision of urban services, political incorporation, everyday life, and urban planning and management. These cities serve as a "laboratory" for observing patterns of the embedding of diversity in urban structures and institutions.

A multicultural city evolves with the recognition and accommodation of cultural differences in the norms, practices, and policies of a city's institutions. As Will Kymlicka points out, "The policies that are intended to recognize and accommodate the distinctive identities and aspirations of ethno cultural groups" are the basis of multiculturalism.[7] A multicultural city is not a new town, evolving fresh out of the ground. Rather, the recognition and accommodation of diverse identities and cultural practices into a city's existing structures and institutions turn it into a multicultural city.

An equally important question is how multicultural cities cohere and function as civic societies, accommodating differences and integrating diverse cultures. What makes these cities vibrant as well as cohesive places? This book, through its two-sided model of multiculturalism (to be elaborated in the next chapter), examines and explains the integrating processes, building on the concept of a common ground. How does differentiation work in tandem with processes of integration in making a multicultural city? This question will also be pursued in analysing the incorporation of diversity into the institutions and structures of the three cities. This is largely a new approach and a unique contribution to the study of multicultural cities.

This is essentially an empirical work based on concrete evidence. For example, accounts of Chinese economies, Latino enclaves, culturally appropriate provisions of urban services, the incorporation of minorities into political representation, and the sociology of everyday encounters across ethno-cultural differences in the three cities form the bulk of the material in the book. These accounts are informed by theories of both multiculturalism and urban structures, articulated in accessible language. The narrative is meant to be concrete and evidence-based.

To sum up, the book aims at offering:

- A generic description of multicultural cities, based on observations of Toronto, New York, and Los Angeles, identifying patterns of their evolving institutions and structures under the influence of multiculturalism.
- A comprehensive account of the infusion of a multicultural ethos and pluralism in the urban institutions based on the experiences of the three cities.
- Identification and explanation of the similarities and differences in the multicultural norms, practices, and policies of the three cities.
- An analysis of the processes of differentiation and mechanisms of integration that lead to the evolution of common ground, including the role of civic culture and urban space as integrating mechanisms.
- Observations about the limits of cultural diversity and the role of a common ground.

Multicultural Cities as a Conceptual Problem

The core conceptual issue of multicultural cities is how they accommodate and recognize diverse cultures and identities, of migrants in particular, and how they function as a cohesive and unified urban society of shared citizenship. The city exists before the arrival of immigrants that bring new identities and cultures. Its history, geography, economic organization, cultural traditions, and social structure, as well as political order, are long established. Into these structures and institutions are injected diverse ethno-racial identities and different values, norms, and practices, legitimized by the contemporary regimes of human and civil rights.

The demands of democracy and equity require that people's needs are appropriately satisfied in culturally and socially responsive ways. The goal of equity in this case relates to fairness in the enactment of cultures and identities, beyond economic and political matters. It requires pluralistic norms and policies. But these diverse norms and practices have to be integrated into a unified civic culture and common values. This dynamic of cultural and social differentiation and integration is the challenge of studying multicultural cities. Ash Amin alludes to this aspect of multicultural cities in stating that in a democratic multi-ethnic society "the key challenge is to strike a balance between the cultural autonomy and social solidarity."[8]

The concept of the multicultural city is seldom explicitly defined, though it is extensively used.[9] The discussion of multicultural cities overlaps with narratives of immigrants' contributions to cities and problems in their integration. It also veers towards questions of ethnic relations and racial inequities. These topics have a bearing on the multiculturalism of cities, but they are not the core issues. Cultures and social organizations are in the foreground of this inquiry. What characteristics make a multicultural city? This is the question at the heart of the discourse about multicultural cities.

Toronto, New York, and Los Angeles as Multicultural Cities

To observe empirically the defining characteristics of multicultural cities, I will examine the effects of multiculturalism on the structures and institutions of three North American cities, Toronto, New York, and Los Angeles, though occasional references will be made to other cities for comparison.

Toronto is a Canadian city that embodies the national policy of multiculturalism. New York and Los Angeles are American cities with long traditions of immigration and pluralism. Altogether, the three cities, with their similarities and differences, are a good "laboratory" to observe how cultural diversity affects urban structures and processes. Comparative analysis of their spatial, economic, and social institutions will help bring out the conditions necessary for infusing a multicultural ethos into cities.

The analysis will focus on both the primary city and the metropolitan areas of these cities. According to the most recent census, Toronto city had a population of 2.62 million in 2011. Its census metropolitan area (CMA) population was 5.58 million, which included Toronto city and twenty-two other municipalities as well as an aboriginals' territory. The CMA is different from the Greater Toronto Area (GTA), which refers to an administrative alliance of local and regional municipalities overlapping but not limited to the CMA.

New York City had a population of 8.18 million in 2010. By itself it is large enough to be a metropolis. Its census metropolitan area population of about 19 million extends over New York City, Connecticut, northern New Jersey, and Long Island, too spread out a region, spilling across three states, to be a useful basis for comparative observations.

Los Angeles city had a population of 3.8 million in 2010. It is the centre of Los Angeles County, whose population of 9.82 million (2010) is linked in economic and spatial bonds, which makes the county the functional metropolitan region. The Los Angeles census metropolitan area extends to five counties and has a population of 18.2 million and an area of 33,400 square miles, which is too large and institutionally diffused to be an appropriate metropolitan region for this study.

The central cities and many surrounding municipalities of the three cities are now majority-minority places, where no single ethno-racial group is in majority and a combination of minorities make the majority. I will examine in detail the demographic composition of the three cities in chapter 3. Presently, I will give a thumbnail sketch of their ethno-racial diversity.

Toronto is increasingly a city of immigrants, 48.6% of its population was foreign-born, as per the 2011 National Household Survey (NHA). Equally striking is the fact that the visible minorities (the Canadian term for non-Whites) formed 49.1% of the total population. For the Toronto CMA, the percentages were almost the same: 46% immigrants and 47% visible minorities, with European-origin Whites at the cusp of being a minority, as almost 70,000 immigrants stream into the metropolitan region every year. Chinese, South Asian, Filipinos, Italians, and Jews were the leading ethnicities. This ethno-racial transformation has happened in the last thirty years.

In New York City, non-Hispanic Whites formed only 33.3% of the population in 2010, followed by Latinos at 28.6%, Blacks at 25.6%, and Asians at 12.7%. Immigration is fast diversifying the city's population. Asians and Latinos are the fastest-growing ethnic groups.

Hispanics/Latinos were the largest ethno-racial group in both Los Angeles city (48.5%) and the county (47.7%), as per the national census in 2010. Anglos and other non-Hispanic Whites were, respectively, 28.7% and 27.8%, a minority. In Los Angeles city as well as the county, Blacks have been historically a small minority, but their percentage in the city has declined from 11.3 to 9.7%, and in the county from 9.8 to 8.7%, between 2000 and 2010, largely because of immigration.

The point is that these three cities are demographically and culturally turning into a new genre of urban places, where minorities are becoming majorities and ethno-racial diversity is the structural condition. Immigration is the driver of this change. These cities present an ideal field for the study of multiculturalism's urban impact.

Why These Three Cities?

Toronto, New York, and Los Angeles are the largest metropolises of Canada and the United States respectively. They are also the premier gateways for immigrants, with long histories of accommodating multi-racial and multicultural populations. Although each city has a unique history and socio-political organization, they have many parallel ways of incorporating pluralism in their institutions.

They are acknowledged multicultural cities (as noted by Glazer, Foner, and others) and are hubs of international finance and trade as well as art and entertainment. Their policies and practices of managing ethnic and racial differences are followed by other cities. In this respect they are the "laboratories" of forging multicultural urban institutions. A comparative study of their multicultural policies and practices is a social experiment that highlights the similarities and differences in their approaches and identifies conditions that help transform cities into multicultural societies.

Finally, the dynamics of diversity of these three cities has been exten-sively documented. The material available on their management of diversity is unprecedented. This resource makes the task of this book manageable. It should be pointed out that the book is not about the development and structuring of these cities, but about the incorpora-tion of a multicultural ethos in these urban societies. On this score, these cities are fruitful sites for study.

The Cultures of Cities and Theoretical Discourse

Ethno-racial diversity spawns the multiplicity of cultures and lays the basis of demands for the recognition of the identities of their bearers. This brings up the question of the cultures of a city and what are their bases.

Theoretical interest in the culture or cultures of cities has a long his-tory. At the dawn of the modern period, theorists such as Max Weber, Emile Durkheim, and Georg Simmel were occupied with exploring the cultural elements of city life. The Chicago School of urban sociology in the early twentieth century identified urbanism as a distinct way of life and urban social organization as the embodiment of its culture and social relations. Urban theory continues to command interest in the cul-ture of cities right up to the present time, as reflected in the writings of David Harvey, Manuel Castells, Anthony Giddens, and the purveyors of the theory of postmodern urbanism, sometimes called the Los Angeles school of urbanism, namely, Michael Dear, Edward Soja, Allen Scott, and others. What theoretical threads can be picked up from these writings will be discussed later. At this point, my focus is on the idea of "culture" embedded in the notion of multiculturalism.

Anthropologically, culture has meant the way of life of a society, including its language, religion, beliefs, values, customs, art, symbols, and so on. It is also defined as a system of signifying symbols and images that is enacted in everyday life. Yet the term *culture* is neither restricted to societies or nations, nor is it meant to be a complete code of life.

Culture as a shared mental and symbolic map of behaviours, as a network of representations that shape social life,[10] has also been associated with social groups smaller than a society. UNESCO defines culture as "a set of distinctive spiritual, material, intellectual and emotional features of society or a social group ... It includes art, literature, ways of living, value system, traditions and beliefs."[11] Note the reference to a social group also as the bearer of culture.

Within societies, folk, rural, youth, urban, high or low cultures have been recognized to be different, as are almost invariably ethnic cultures. In these days of corporations' domination, the concept of organizational or corporate culture is widely used. This "culture" refers to a set of norms, values, symbols, ideologies, and practices that guide the behaviours of members of such groups in the corporate sphere. All of these are cultures of limited scope. They are subcultures.

A subculture is organized around a set of values, distinct behaviours, and shared customs that a number of people follow in associating with each other and in expressing their shared identity.[12] It may not include legal, political, technological, or economic behaviours. It is in this sense that the "cultures" that constitute multiculturalism are to be viewed. The Latino, European, or Asian cultures found in cities are subcultures. They are not territorial cultures of separate nations or societies.

Ethno-racial subculture in a society includes the language, religion, norms of domestic and family life, identity, music, food, literature, customs, and festivals of a group. Yet it seldom includes the beliefs and behaviours of workplaces (except for those in ethnic economies), travel, politics, citizenship, law, administration, public places, markets, schools and universities, and so on. The point is that the "culture" of multiculturalism provides the blueprint for some aspects of life and not for others that are regulated by other cultural systems.

An Arab resident of Toronto, for example, speaks Arabic at home, socializes largely with relatives and friends of similar backgrounds, prays in a mosque, lives in an area where there are many Arab families, stores, and restaurants, listens to Ferouze's or Um Kulsom's songs, and is a practising doctor. Yet this is not her total life. She shops downtown, works in a public hospital, aspires to send her children to Harvard or Cambridge (but would like them to marry Arabs), vacations in the Caribbean, and votes for the Liberal Party. Her way of life is a mixture of an Arab subculture and the societal culture of Canada. Her Arab subculture is the culture of a community or group that is part of the city of Toronto and Canadian society. Similarly, Chinese or Jewish cultures, for example, represent only segments of the daily life of such groups in particular cities.

The scope of these subcultures varies from group to group. Some, like American Blacks, may have the same language and religion as the mainstream but a distinct historical experience and identity, while Indians and Pakistanis have different religions, languages, customs, and codes of behaviour. Whatever their scope, they do not offer a total way of life.

There are other cultures operating at different levels in a city. There is the culture of behaviours in public spaces. As Sharon Zukin says, public culture is a source of images and memories that symbolizes "who belongs in specific places."[13] It regulates encounters among people of diverse interests and expectations.

Civic culture is another term used to describe the codes of behaviours applied to the socio-political arena of social life. The term, coined by Gabriel Almond and Sidney Verba, postulates a form of social contract in which citizens accept the authority of the state and are afforded participation in civic affairs.[14] If one extends this political formulation of civic culture to social behaviours, it can be defined as norms of behaviours in public places, municipal and regional laws, and mores of participation in services, images, memories, and local spirit.[15] Civic culture includes but is not limited to the public culture of a city.

Thus, in a city there are three layers of culture: (1) ethnic, (2) civic, and (3) societal. Boundaries between these various cultures are soft and they are continually borrowing from each other. All in all, the dynamism of cultures must be kept in mind in the study of multicultural cities. Yet the idea that cultures of multiculturalism are subcultures, limited in scope, is the foundational concept of this book. Claude Fisher makes subcultures the structural feature of urbanism. He concludes that "urbanism probably promotes the emergence of numerous and diverse subcultures within a community."[16]

The Scope of Community Culture

The carrier of a culture of multiculturalism is a community or group. As discussed above, the domain of community culture is largely the traditions and norms of family, friendship, and domestic life as well as the patterns of religious, artistic, and associational behaviours of an ethnic group, namely, the private domain. The community culture is inward looking and tied with the identity of a group and its members.

Community is also, like culture, an elusive concept. It is widely used to describe social organizations or constellations of persons or groups linked together by common social bonds, interests, territory, memory, myths, and/or history. It also intimates the quality of social relations, namely, warm and cooperative. In this postmodern era, electronic means of communication and accelerated social mobility have liberated the community from territorial bounds. Now social relations and shared interests can be sustained across dispersed locations. Propinquity is not necessary for the formation of a community.

The community now lies in overlapping social networks that link people around shared interests and mutual exchanges. Barry Wellman and Barry Leighton classify communities into those of (1) dense, (2) ramified, and (3) sparse networks.[17] Normally, in the context of a city or country, a community refers to the intermediate-level social organization. This is how I am using the term.

In cities, communities are formed around common origins and roots, languages, religions, and traditions, albeit elements of the ethno-racial identity, as well as on the bases of shared lifestyles, professions, values, and interests. Community sentiments that cut across these bonds also emerge around territorial neighbourhood and other local interests. Communities are not mutually exclusive but overlapping. An individual is simultaneously a member of many communities.

From the multicultural perspective, ethnic communities are the most significant cultural groups, though lifestyle and shared-values communities also have distinct subcultures. When we talk about multiculturalism, we are primarily talking of the cultures of Italian, German, Chinese, Mexican, Indian, or Aboriginal communities in New York, London, Toronto, or Sydney, for example. Normally, they are not self-contained, but are embedded as patches in a mosaic. Communities are subject to the common civil and criminal laws, economic institutions, public ethics, and national ideologies. These are matters beyond the bounds of community culture.

Like all cultures, community cultures are not monolithic and fixed in space and time. They are internally diverse, based on a range of norms and customs, and evolving over time, particularly when they exist side by side with other cultures.

Community cultures primarily arise from the transplantation of immigrants/in-migrants in cities. They usually refer to minority sub-cultures implanted in the mainstream culture of the historic majority. With increasing enactments of the protection of minority group rights internationally, community cultures have gained renewed legitimacy.[18] They are giving rise to new conceptions of the culture of a society and nation. It is increasingly conceived as a federation of subcultures and the society or city itself being thought of as a "community of communities."

Before concluding this discussion of the community culture, it should be pointed out that ethnicity is a predominant base of cultural diversity, but it is not the only source. Lifestyles, social values, and identity politics are other contributors to the cultural diversity of cities and societies. Bohemians, Yuppies, Punks, or, to use David Brooks's term for the new upper class, Bobos[19] are often viewed as distinct cultural communities on the basis of their lifestyles and values. Yet much of the multicultural-ism discourse is about ethnic communities and their subcultures. This is how the term community culture will be used in the book.

Ethnicity, Identity, and Community Culture

Ethnicity is a social boundary that defines who is inside and also, by implication, outside a group. The boundary is based on a shared culture, language, history, beliefs, religion, and/or origins and it often includes racial and physical differentiations.[20] One is Anglo-Saxon, German, or Japanese on the basis of one's heritage, culture, national-ity, geographic origins, language, and often race as a member of some imagined group.[21] These nationalities transplanted in Canada or the United States turn into ethnic groups and form the bases of community cultures.

Yet ethnicity is a social construct. It varies with the context. I am a Punjabi in Pakistan, but a Pakistani in Britain, but along with Indians and Bangladeshis become a South Asian in Canada. Puerto Ricans or Mexicans have distinct ethnicities in the United States, but are swept

into the broad community of Latinos. A similar context-driven redraw-
ing of ethnic boundaries takes place in cities, resulting in the alignment
of community cultures. African-Caribbeans come to New York and
become Blacks, but in London they are West Indians.

Another element that underlies community cultures is the sense of
identity. One's sense of self derives from one's age, physique, gender,
social status, nationality, ethnicity, and occupation. Anthony Appiah
divides identity into two dimensions, namely, collective and individual, but only the collective identity (race, ethnicity, religion, etc.) counts
as a social category.[22] Recognition of identity is a part of an individual's
or a group's authenticity.

Everyone has multiple identities. Amartya Sen maintains that the
presumption that a person belongs to one group only is unjustified.[23] In
the context of city living, often the identity derived from ethnicity and
race has a strong traction. This is particularly true for immigrants. Even
in the second generation, one's ethnic identity continues to be a defining element. Many US- or Canadian-born second-generation Chinese or
Latinos, on being asked "When did you come to Canada/the US, you
speak such good English," discover that their ethnic and racial identity
defines them more than their birth, citizenship, and even profession.

The interplay of ethnicity, identity, and community culture leads to
the multiplicity of cultures in cities. It weaves into racial differences and
gives rise to the discourse about diversity, equity, and citizenship. Community identities affect an individual's and a group's status in a city.

Religion and Community Cultures

The academic discourse on multiculturalism has largely ignored the
role of religion in defining community cultures and identities. Often
religion is looked upon as a matter of the individual's right of freedom
to believe and practise. The state largely keeps out of this area, except
to protect the right to religion and protect itself from the intrusion
of religion into its affairs. This is the secularist bias in the world view of
the social sciences.

Yet religion is a salient element of the cultural landscape in modern
societies. Religious customs and ethics are enacted in the public sphere.
They are inseparable from culture. Witness the present controversies
about same-sex marriage and abortion in conservative ideologies and
the recurring themes of Muslim and orthodox Jewish women's self-
segregationist practices as issues of public culture.

Since the late twentieth century, there has been a revival of religion both nationally and internationally. Islamic fundamentalism is paralleled by the Christian, Jewish, and Hindu revivalist movements. The popularity of Buddhism among the secularized segments of Western societies is another indication of the increasing role of religious values in structuring communities. All these examples are meant to argue that (some) religious practices and morals are woven into cultures of communities and societies.

The religions of the current wave of immigrants in Europe and North America, particularly of those coming from Asia and Africa, fall outside the historical Christian faith. Islam, Hinduism, Sikhism, and Buddhism are religions at variance with the Judeo-Christian traditions of Western societies. They underlie certain cultural differences surfacing in the United States, Canada, Britain, and France. Many cultural controversies are rooted in religious differences. Religion is also a strong element of people's identity and community feelings.

The banning of religious symbols in France's public schools was prompted by the public controversy about Muslim girls wearing head coverings (hijabs). Recently, the niqab, or face-veil, has precipitated heated public debates in Britain, France, and Canada. Public debates about women's status in North America and Europe bring out differences of religious values, which turn into cultural clashes. Similarly, the Sikhs' turban is incompatible with the safety requirement of wearing a helmet for bicycle and motorcycle riders. These examples point out that cultural diversity is also complemented by differences of religious values and customs. They suggest that community cultures include some religious values and practices. How religious differences intertwine with cultural practices and ethics in the public space is another factor in determining the scope of multiculturalism.

Narratives of Differences

The discourse of multiculturalism is largely a narrative of cultural differences. Popularly it takes one of two forms, though the scholarly literature is much more nuanced. Either it is celebratory in tone, pointing out the quaint customs of different communities, or it flags the risks of social divisions and ethnic strife arising from cultural and racial differences. The former celebratory theme tends to emphasize the exotic elements, recognizing the equality of community cultures in the mosaic making up a societal or urban culture. The latter theme draws on the apprehensions of minorities' separatism and lack of social cohesion. In the literature on diversity in Western cities, the theme that minorities and immigrants are victims of discrimination also resonates strongly in liberal-left formulations. All in all, cultural differences remain the focal point of national conversations on multiculturalism.

The normative expectation of equality in liberal democratic polities not only calls for the acceptance of subcultural differences but also necessitates conceiving the national society as a composite rather than a unitary culture.

How are cultural differences negotiated in a city and how does it cohere? This question has not received adequate attention in the multicultural discourse. For the time being, it is enough to point out that cultural differences within a society coexist with the common ground of institutions, laws, and ideologies linking all groups together. The critical questions are how does this common ground develop in an urban society and how can a city help forge the common ground? They will be taken up in the next chapter.

What Next?

In this introductory chapter, the scope of the cultures constituting multiculturalism has been explored. The culture of multiculturalism is limited. It is embedded in the civic culture of a city and in the overarching societal culture. The chapter lays out the definitions of basic terms, that is, culture, community, ethnicity, and identity. These definitions will frame further discussions. Chapter 2 is an attempt to clarify and spell out the concepts that underlie the observations and analysis of subsequent chapters. It offers an examination of the concept of multiculturalism, its various forms, and political and social discourse about diversity and social cohesion. The idea of common ground is examined as the complement of multiple cultures. How these concepts relate to the ideas of citizenship and rights to the city[24] forms another theme in this chapter. Chapter 3 explores how immigration lays the basis of multiculturalism and examines its role in implanting cultural diversity. Chapter 4 is an analysis of the spatial structure of multicultural cities. Chapter 5 describes the ethnic contours of the economic organization of multicultural cities in the three cities and explores the phenomenon of ethnic entrepreneurship in urban economies. Chapter 6 analyses the social organization of multicultural cities and their constituent communities. Chapter 7 recapitulates the experience of everyday living in multicultural cities. Chapter 8 is a description of the trends in the political incorporation of ethnic communities. Chapter 9 examines the process of accommodating cultural diversity in the provision of urban services. Chapter 10 analyses the state of planning for multicultural cities and develops an approach to balance cultural diversity and public interest. Finally, chapter 11 sums up the findings of previous chapters and develops a generic model of a multicultural city.

Cultural diversity and its transforming influences on the social, economic, political, and spatial institutions of a city are the focus of this book. Immigrants' settlement and integration are not its primary focus, though their roles as bearers of cultural diversity are analysed.

This book lies at the intersection of two theoretical discourses. It is firmly planted in the discourse on multiculturalism, on the one hand, and in urban theories of the spatial, economic, and social and political/managerial institutions of cities, on the other. The former discourse will be explored in the next chapter and then the application of multiculturalism will be examined through the theoretical lenses of the respective topics in subsequent chapters. The focus remains on the infusion of cultural pluralism into urban institutions.

Another point to be clarified is that there are vast and distinct bodies of literature on the spatial, economic, social, political, and policymaking aspects of cities in general, but not on the three cities in particular. I will draw selectively on relevant disciplines, theories, and empirical observations in subsequent chapters, guided by my focus on the infusion of multicultural ethos into each institution. Obviously, the book will not attempt to summarize the vast body of literature on each topic and about each city. Such a task is neither necessary nor feasible. Instead, it will recapitulate those theories and observations that have direct bearing on the objectives of the book.

Chapter Two

Multiculturalism: Diversity Rights and the Common Ground

Multiculturalism

A study of multicultural cities draws on two theoretical discourses: (1) the ideas of multiculturalism and (2) the theories of the city as the geographic base of economic, political, and social institutions. In this chapter, I will critically review the notion of multiculturalism and the ideas about cities, particularly through the lens of their cultures and social organization. My objective in reviewing these concepts and theories is to develop indicators and categories by which I will observe and analyse the multicultural dimensions of the three cities. I begin with the question, What is meant by multiculturalism?

Multiculturalism is a term in use with a wide range of meanings, particularly in the popular media. Often immigrants' practices and customs are described as synonymous with multiculturalism. If a Muslim woman puts on a niqab, or face-veil, the practice is attributed to multiculturalism, even though it may be an individual act. Incidents of terrorism in Europe and North America have been explained as the outcome of laxity in immigrants' integration due to the policies of multiculturalism. On the positive side, multiculturalism is identified with the diversity of foods and music as well as the creativity of the global cast of talent in North American cities. These images of multiculturalism are partial pictures, as the following discussion will show.

Among academics, multiculturalism sparks debates about the nature of political community that it promotes, or leads to arguments about its liberal, conservative, or communitarian intents. Arguments rage over the rights of ethnic and racial minorities; apprehensions are voiced on perpetuating social segregation and eroding national solidarity. All in all, there is little consensus on the scope and limits of multiculturalism, though there is some unanimity about its field of operation or domain. These arguments call for clarifying the concept of multiculturalism. I will begin with a definition.

Multiculturalism is a social condition, political ideology, public policy, or project that, occurring in various combinations, helps realize the recognition and expression of cultural and ethnic diversity within a nation or its parts. Obviously, the point of reference in multiculturalism is not the multiplicity of national cultures in the world. It is the cultural pluralism within nation states and in defined spaces of cities and regions. Multiculturalism is also an ideology of recognizing the rights of ethnic, racial, linguistic, and religious minorities to express their distinct identities and cultures freely. These rights apply at both the community and individual levels.

I have already discussed the nature of "cultures" in the context of multiculturalism in the previous chapter. Here I will extend the theoretical and analytical discussion of multiculturalism as a foundational concept. This chapter also develops the idea of common ground as a component of multiculturalism. It shows how common ground is necessary for linking together diverse cultures and identities in the shared space of a city as well as a civic society. These concepts provide the framework for examining changes in cities' structures and institutions brought about by the incorporation of diversity. Furthermore, the chapter links the notion of multiculturalism with the ideas of cities, citizenship, and the right to the city.

Multiculturalism is a conception of society in which several cultural and ethnic groups are recognized to be equal in rights and identities. It embodies the goal of living with diversity to promote "equality, justice, and an expanded level of societal solidarity."[1] As the focus of this book is on how multiculturalism as a phenomenon transforms the structure and institutions of cities, it is appropriate to begin with a sociological understanding of its scope.

British sociologist John Rex divides social life into two domains, public and private. The main institutions of the public domain are law, politics, and economy,[2] though others such as technology and civic order can also be included. The private domain is structured around family, kinship, religion, community, and identity. Each domain has its norms, values, moral order, and ethical codes. This differentiation of the two domains has been widely used to define multiculturalism.

Rex identifies four ways in which the public and private domains can be combined. A unitary public domain combined with the diverse private domain makes up multiculturalism. While both domains being unitary would represent the ideology of assimilation or the melting pot, diversity in the public domain (different laws for different communities) combined with the diversity of the private domain represents the situation of colonialism, pre-independence South Africa, or Israel, for example. To complete the fourfold combinations, a diverse public domain could be conceived to exist with a unitary private domain, a situation that Rex identifies as similar to the pre-civil-rights US Deep South.[3]

As the first country to pass a national multiculturalism policy (1971), Canada acknowledged "the freedom of all members of Canadian society to preserve, enhance and share their cultural heritage ... to ensure all individuals receive equal treatment and equal protection under the law, while *respecting their diversity ... to encourage and assist the social, cultural, economic and political institutions of Canada to be both respectful and inclusive of Canada's multicultural character*" (emphasis provided).[4] In this act, the inclusiveness of the Canadian public institutions is backed up with the recognition of the cultural rights of individuals and communities. Will Kymlicka holds that multiculturalism combines "robust forms of nation-building with robust forms of minority rights."[5] The Parekh Report of the Runnymede Trust in Britain also holds a "single political culture in the public sphere and substantial diversity in the private lives of individuals and communities" as the liberal model of multiculturalism.[6] The common cultural elements of the public domain are as important as the differences in norms and values of the private domain for a multicultural society to function.

This is the theory. Yet in practice, the relationship between the two domains is not as clear cut as the above description may suggest. Their respective spheres overlap and boundaries are fuzzy, shifting with political and social changes and giving rise to various forms of multiculturalism.

Fuzzy Boundaries and a Shifting Balance

The boundaries between the two domains are permeable. For example, the child rearing that unquestionably falls in the private domain is subject to the child welfare laws of the public domain. In the United States and Canada, a parent is not free to discipline children by beating them.

Similarly, beliefs and practices of the private domain rise into the public domain with changing ideologies; for instance, same-sex marriage as a private practice is now entering the public arena in the form of gay marriage laws.

The point I am trying to drive home is that the boundaries between a uniform public domain and a diverse private domain are permeable. They shift as a part of social change and political development. The civic culture based on the values of equality, fairness, and democracy, and on the rights and responsibilities of citizenship, is the bridge between the two domains.[7]

Ultimately, multiculturalism is not just a matter of the recognition of diverse communities living by their cultures in their private life, but a process of integrating their interests and shared values in the public sphere.

Theoretical Perspectives

Theoretical discussions of multiculturalism are largely framed by political debates about the preferred polity for multi-ethnic and multiracial societies. They shift with changing political ideologies and interests. There is a large body of literature on the variety of forms multiculturalism takes, its critiques, and counterarguments. As the focus of this chapter is on clarifying the concept of multiculturalism and identifying its indicators to observe the institutionalization of multiculturalism in the three chosen cities, I have summarized the theoretical discourse of multiculturalism in the Appendix for readers interested in such arguments. For the task at hand, namely, the spelling out of a workable model of multiculturalism, the following discussion will lay out the premises and elements of testable propositions.

Many theoretical formulations are conceived as responses to the political critiques of multiculturalism. This explains why, as Peter Kivisto says, most of the theories and philosophies of multiculturalism are normative and not based on empirical findings.[8] Though "culture" is at the core of the issue of multiculturalism, there is relatively little sociological and anthropological theorizing about multiculturalism.

Augie Fleras identifies twenty-two conceptual questions on multiculturalism, including how can diverse communities live together with common values; does equality mean treating everyone exactly the same or differently; how is the tension between culture and social justice resolved; and why is the politics of culture privileged over social integration?[9] These are some of the questions around which multicul-turalism's theories have evolved. By and large, most questions can be subsumed under three themes: (1) the criteria for recognizing the diversity of groups and individuals, be it ethnicity, culture, race, identity, religion, or a combination of various attributes, (2) balancing the rights and claims of diverse groups with the demands of national integration and common values of citizenship, and (3) weaving diverse identities and interests into the common public institutions and the reimagining of the societal mainstream. Although the theorists of multiculturalism (Kymlicka, Fleras, Taylor, Parekh, Madood, Banting, Kivisto, and so on) differ from each other in interpreting these themes, there is a general convergence of their views towards multiculturalism's promise of sustaining both diversity rights and social cohesion.

Much of the theorizing about multiculturalism happens in Canada, Britain, and Australia, the countries that have adopted multiculturalism as a national policy. Therefore, the theories largely reflect the assumptions and concerns of implementing multiculturalism. Framing the discussion around the three themes, described above, helps in reviewing the theoretical formulations.

First, multiculturalism privileges culture as the basis of recognizing and accommodating differences, but the culture is attached to ethnicity, race, or religion as identities. The criteria for defining diversity rights are "distinctive identities and aspiration of ethno cultural groups," as per Kymlicka, which wraps together both culture and ethnicity.[10] Charles Taylor lays emphasis on identity as a "person's understanding of who they are" and stresses the need for recognition at the levels of both self and public.[11] He argues for equal dignity for all citizens and a politics of difference. Peter Kivisto comments on Kymlicka's and Taylor's criteria that "the collectivities they have in mind are ethnic."[12] Tariq Madood calls for public recognition of one's heritage as a collectivity along with others and he includes religion as a possible basis of individual and/or group identity.[13] All in all, multiculturalism is primarily about ethnocultural identities, though race is woven into ethnic identities, as the basis of multicultural differences.

The second issue in multiculturalism theories is the balancing of ethnocultural differences with the demands of national integration and the common values of citizenship. Most of the criticisms of multiculturalism are based on questioning the feasibility of forging a coherent society by recognizing and entrenching differences. Kymlicka maintains that the recognition and accommodation of ethnocultural differences goes along with the cultivation of "overarching national identities and loy-alties through education, citizenship, language, funding of media and history."[14] And he says that "multiculturalism combines robust forms of nation-building with robust form of minority rights."[15] Augie Fleras argues that multiculturalism is "better positioned to achieve social cohesion, economic integration, secure positive identities, meaning-ful citizenship and avoid ethnic strife."[16] John Rex observes that mul-ticulturalism "involves acceptance of a single culture and a single set of individual rights in the public domain and variety of folk cultures in the private domestic and communal domain."[17] These theorists see multiculturalism as the policy that reconciles cultural diversity with the unity of societal values and norms.

The third theme is about the structure of common public institutions and the reimagining of the mainstream. Again to quote Kymlicka, "The purpose of multiculturalism is to renegotiate the terms of integration."[18] The public institutions in multiculturalism are meant to incorporate the interests of minorities and reflect values of pluralism. Their restructur-ing is as much a part of the process of multiculturalism as the recog-nition and accommodation of diversity. Keith Banting and colleagues posit the idea of shared citizenship as a reformed public institution that realizes "a certain level of social integration and engagement with the broader community."[19] Bhikhu Parekh envisages the public realm as open to revision and as respecting different cultural identities.[20] Call it what you may, a common civic culture, renegotiated public institutions or shared citizenship, the restructuring of the national narrative, and the reimagining of the mainstream are integral parts of multicultural-ism. Institutionally, a common ground reflective of minorities' interests is necessary to sustain the ethnocultural diversity in multiculturalism.

Multiculturalism Discourse in the United States

The United States' national narrative is that it is a melting-pot country and a land where assimilation of immigrants is the rule. Yet ethnic and racial identities have long defined the social contours of American soci-ety. The idea that cultural pluralism is a defining characteristic of Amer-ican society has been a persistent theme in social commentaries and academic discourses. Horace Kallen held it to be the American condition as early as 1924.[21] In a later section, I will examine the state of assimila-tion in the United States. For now it is enough to note that ethno-racial communities have thrived in the United States, and the social structure is etched with their identities and differences. Witness the long history of the segregation of Blacks. Also the cultural differences among Irish, Italians, Jews, and Eastern Europeans, among others, are woven into the mainstream. On the ground, the United States has been a society of multiple cultures, with the difference that its national identity and pride in its exceptionalism (manifest destiny in early times) are stronger than other settler societies like Canada and Australia.

American multiculturalism is primarily lodged in educational policies, curricula, employment equity, politics, and ethnic imprints on cities and regions. It does not come much into the discourse about the national polity. It is largely a matter of individual and communal initiatives in the spirit of liberty and free enterprise. How the discourse about multiculturalism has evolved can be observed from the writings of Nathan Glazer, a magisterial scholar of ethnic and race relations and education.

In their 1963 book *Beyond the Melting Pot* Nathan Glazer and Daniel Moynihan found that ethnicity and religion were important for politics and culture in New York City, and they observed that the newest migrants to the city, Blacks and Puerto Ricans, could be following the path of other ethnic groups.[22] They described New York City's race and ethnic social organization as a form of cultural pluralism, and did not use the term multiculturalism. By the second edition of the book, in 1970, they were recanting their conclusion, observing that religion was in eclipse but ethnicity and race were the major lines of division. Black and White differences were found to be the line of fissure in the social structure. There are other issues raised in the book, but from our perspective, the multiple cultural and racial communities speak for the endurance of diversity in the United States.

In 1991, in a retrospective look at his book in a lecture at the University of Toronto, Glazer conceded that continued immigration had sharpened the ethnocultural differences, which were visible in New York.[23] By 1997, he was proclaiming, "We are all multiculturalists now." He was conceding that a change had swept the educational system and multiculturalism had displaced the old notions of a uniform canon of studies. He concluded "[Let us] respect the diversity of American origins but appreciate fully the power of the integrating values of our common society."[24] The evolution of Glazer's ideas illustrates the realization of how deep the roots of diversity in the United States are and how in this era of cultural rights and the honouring of diversity, multiculturalism has come to the forefront of the American discourse. Now scholars

Photo 2.1 Hair cut in four languages, Brooklyn, New York (courtesy Milagros Dorregaray)

often use the term multiculturalism, particularly in California and New York, to describe the state of ethnocultural composition of the society. Peter Kivisto in a recent article concludes that "there are political and cultural structures in place that will serve to sustain demands made by stigmatized groups for recognition and respect, to the extent that this is true, then we really are all multiculturalists now."[25]

In the United States, multiculturalism is a counter-hegemonic discourse.[26] It is rooted in minorities' assertion of their identities. Many advances in the acceptance of diversity have been made at the national level, yet local and regional accommodations of ethno-racial differences have led the way in instituting multiculturalism. Urban theories point to multiculturalism as the defining characteristic of cities such as New York, Los Angeles, Chicago, and Miami.

The Common Ground

The foregoing conceptual discussions suggest three propositions as the basis of multiculturalism. (1) Cultural differences within a multicultural society are limited to a few areas of social life. (2) Recognition and expression of such differences are enshrined in the civil rights of ethnoracial groups. (3) Shared public institutions, citizenship, and common values are the integrating structures, which incorporate the interests and identities of different groups.

Multiculturalism thus is a two-sided coin. On the one side is the relatively confined culture, rather subculture, of the private domain, and on the other side are institutions of the public domain constituting the common ground that knits together diverse identities and interests into common values, narratives, and symbols as well as shared citizenship.

A common ground draws on a constitution, laws, economy, administration, official language(s), as well as values of a national sweep such as tolerance, freedom, trust, equality, non-discrimination, the rule of law, public welfare, and norms of everyday behaviour. It overlaps with societal culture but differs in that it includes a pluralistic ethos of citizenship and moral codes of everyday behaviours, such as lining up to get on a bus, punctuality, non-violent resolution of disagreements, and cleanliness.[27]

The common ground is embedded in the societal culture but the two are not synonymous. It is a segment of the societal culture, namely, those institutions, values, and norms that regulate the public space and underpin the interactions of diverse groups. There are other elements of the societal culture that have little or no bearing on behaviour in the public space. They could include history, identity, and a state religion among others. Those are normally not a part of the common ground.

The idea of common ground is expressed in different words in many formulations of multiculturalism: in the notions of common values, shared institutions, and integrative policies. Parekh observes that in the public life of multicultural societies, identities are inter-linked and develop common features through the interdependence of communities.[28] The mosaic of community cultures is painted on the common ground.

My use of the term common ground is backed not only by theories of multiculturalism, but also by some historical precedents. The concept of an ethno-racially diverse but inclusive (non-chauvinist) Anglo-Protestant America was advanced in the journal *Common Ground* in the 1940s.[29] This concept can be located in the idea of fraternity, one of the trilogy of national mottos (i.e., liberty, equality, and fraternity) of France inherited from the French Revolution. Ash Amin and Nigel Thrift extend the idea of the commons beyond natural resources to immaterial commons such as health, education, culture, and knowledge. They see them as a part of the right to the city shared by all citizens spanning subcultures.[30]

Shared public space and services forge common norms, values, laws, and ethics to sustain social interactions and a sense of solidarity. In a city, space is an incubator of common ground.

Common ground is not cast in stone, though it has stability and continuity. It evolves with changing conditions. Also it is cultivated by the engagement of diverse groups in civil-society associations, for instance, professional organizations, labour unions, political clubs, NGOs, and parent-teacher associations. These are examples of Robert Putnam's bridging social capital.[31] Ashutosh Varshney in a path-breaking study of ethnic conflicts in India found that cities in which Hindus and Muslims were formally linked with each other in civic and economic associations had kept ethnic peace even in the face of communal tensions.[32] They had built a common ground of mutual trust, which proved to be a strong defence against ethnic riots. There is a lesson here for the imperative of a common ground in a city.

Minorities should see themselves reflected in the common ground in order to be bound by it. Multiple bonds among ethnic communities should reinforce these institutional accommodations. Multicultural-ism is as much a matter of ethnic minorities having the right to live by their culture in the private domain as a one of reconstructing the public domain to accommodate their interests.

The Construction and Reconstruction of the Common Ground

There are two distinct segments of the common ground. One consists of a constitution, laws, policies, conventions, and symbols that lay down the rights and responsibilities of citizenship and the organizing principles of state and society. The second segment consists of institutions, values, morals, and practices that regulate people's beliefs and behaviours in the public space. The former are constructed legislatively, administratively, and politically; the latter arise from socio-political processes of social interactions, civic engagement, educational and media discourses, as well as economic and technological developments. The former elements are more deliberate, while the latter are diffused. Yet both parts of the common ground change over time, varying in function and meaning more readily than in form and structure.

In liberal democratic societies, the process of formulating laws and policies increasingly involves citizens' participation. The inclusion of minorities in legislative and political bodies helps integrate their interests into the common ground. To the extent this happens, the ideals of multiculturalism are realized and the common ground comes to represent all segments of a society.

The sociological values and norms that form the second segment of the common ground have a direct bearing on everyday life. They define the baseline of behaviours in the public space. What dress is considered decent in the public arena? What are acceptable behaviours in dealing with persons of different religions and races? How does one behave on roads and streets? Is a landlord at liberty not to rent her house to a gay couple? Can a school disallow a Muslim girl from wearing the hijab? These are examples of normative questions that arise in a society of diverse communities. Such norms have to accommodate minorities' rights and entitlements, on the one hand, and acculturate them, particularly immigrants, in the expected social behaviours, on the other. This is the process of reconstructing the common ground.

All in all, construction and reconstruction of this aspect of the common ground is an incremental process. In modern societies, the mass media and social movements are the instruments of raising public consciousness about the need for change in values and norms. These sociological elements of the common ground evolve with the injection of mores and practices. A few examples will illustrate this point.

In contemporary societies, national mores of food and dress are affected by contacts with other cultures globally as well as locally. The notion of decent dress in public has undergone considerable change since the 1960s. The protest movements of the 1960s breached these mores and the youth and ethnic cultures as well as feminism have vastly extended the range of acceptable dress such as Capri pants, stretched shirts, halter tops, and suits worn without neckties.

Another example of change in public mores is provided by the acceptance of hugging as a form of greeting. In the 1960s, it was almost a taboo for men to touch each other except for a handshake in North America. In recent times, presidents have been freely hugging visitors and citizens whom they meet in their tours. Amy Best, a sociologist at George Mason University, has said that teenage embraces are a reflection of the evolution of the American greeting.[33]

Changes in food, dress, and public manners may be dismissed as trivial, but they represent the essence of everyday life and symbolize the national identity. Their change is indicative of changing behaviours in the public space. It is an ongoing process, but immigration and the presence of ethnic minorities quicken its pace. This is how common ground is reconstructed in incorporating communal mores. Multiculturalism both accelerates this process and necessitates a more organized accommodation of diversity.

All in all, in a multicultural society both parts of the common ground, laws/policies versus norms/values, have to reflect the interests of cultural minorities. This process happens incrementally in liberal democratic societies. It can be put on a more formal footing by involving minorities also in policymaking and national conversations and by promoting inclusive civic engagement and education.

A common ground reflective of diverse interests serves two purposes. One, it ties together different communities in a society through shared legal, administrative, and moral institutions. Two, it induces minorities to own the societal values and norms and develop a sense of belonging.

As the later chapters will show, the forging of a common ground is an elemental process of multicultural cities. The diversity of socio-cultural life in multicultural cities is sustained by a common ground that ties together the urban structure spatially, economically, and socially.

Assimilation, Integration, and Reasonable Accommodation

The discourse of multiculturalism also brings up other modes of structuring a diverse society. Concepts such as assimilation, integration, social cohesion, and reasonable accommodation also have a bearing on the ways of reconciling the demands of social solidarity with the need for cultural diversity.

On the surface, assimilation and multiculturalism appear to be the opposite of each other. The United States is presented as a society that assimilates minorities and melts their cultural differences, whereas its northern neighbour, Canada, is regarded as a society that officially recognizes ethnic identities and sustains a diversity of ways and values. America as a melting pot is contrasted with the Canadian mosaic. Yet this picture is not a true reflection of the historical as well as the current reality of the two societies. Ethnic identities and customs have flourished in the United States, and Canada has a sense of national solidarity.

Assimilation is popularly thought to be the process of immigrants giving up their identities, languages, and cultures and adopting the ways of the host society. Although this viewpoint has been forcefully espoused by influential Americans, from President Theodore Roosevelt (1904–8) to Pat Buchanan (2008), on the ground assimilation has been a two-way process in which the American mainstream adopted many norms and values of immigrants as they embraced American ways and manners. Richard Alba and Victor Nee have closely examined the process of assimilation and observe that "the blinkered view of the assimilation process overlooked the historical reality that the majority changes too ... and that American society increasingly reflects a composite culture made up of diverse elements."[34]

The ideal of American assimilation was Anglo-conformity, which is to say, Anglo-Saxon Protestantism. It was the model to be adopted, but in fact German cultural influences brought by settlers in the Midwest have been almost as strong an element of American culture as those of the English. The public presence of the German language and culture declined mostly because of the patriotic fervour aroused as a result of Germany being the enemy during the First World War.

Assimilation stopped at the racial boundary. Almost everywhere in the United States, Europe, and Canada, incorporation of non-White minorities failed to occur. They remained socially segregated and culturally distinct. Nathan Glazer concludes that today's assimilation "accommodates more than one identity and more than one loyalty."[35]

It comes closer to multiculturalism. Herbert Gans concurs in this conclusion: "Whatever the pace of political assimilation, individual ethnic identity and ethnic pride will not necessarily disappear."[36]

An assimilating society is not socially and culturally homogeneous. It is divided by class, geography, race, and ethnicity. Immigrants and minorities are assimilated into a specific stratum of society and inducted into its subculture. This is the segmented assimilation theory, which posits that different immigrant groups follow different paths of assimilation depending on the stratum into which they initially settle.[37] This argument points out that assimilation does not lead to homogeneous culture. In reality, its end product is not, metaphorically, the melting pot but the salad bowl.

Academics and policymakers are increasingly using the term integration to describe the incorporation of ethnic and racial minorities, including immigrants, in the political, economic, and social institutions of societies. Integration is a process by which newcomers and others become part of the institutional fabric of a community or society. It allows individuals of one culture to "join in common citizenship with members of other cultural groups."[38] It assumes that "the interests of the various parties will complement one another and add to the development of a civil society without blurring the identities of the participants."[39] James Frideres, a sociologist at the University of Calgary, refers to the three dimensions of integration, namely, (1) social or institutional integration, relating to participation (of immigrants and minorities) in the institutions of a society; (2) cultural integration, leading to the knowledge and observance of common values and beliefs; and (3) identity integration, involving the development of a sense of belonging.[40] Of course, there is another dimension of integration, namely, economic integration through equal access to economic opportunities.

Social integration is achieved by the accommodation of a diversity of private norms in the singular public domain. Tariq Madood defines integration as the process in which both the majority and minority communities make adjustments to forge a public culture.[41] In this process, the reconstruction of the public domain goes hand in hand with the realignment of the private domain.

Integration is a two-way process whose terms are negotiated on an institution by institution basis. Economic integration, for example, means removing racial and other barriers, but in the bargain immigrants and minorities may be required to acquire necessary linguistic and professional skills. Cultural integration may mean reasonable accommodation of minorities' values and norms while they adopt a national/community ethos. Social integration means including racial and ethnic minorities as equals in social networks, communities, and neighbourhoods. It is reflected in fair housing practices. Inclusive social organization's highest form is inter-ethnic and biracial families. Yet in each of these cases, institutions are broadened and realigned to accommodate differences. And that is multiculturalism in action.

A new term has appeared in the realm of social integration. It is social cohesion, and essentially refers to the state of integration of diverse groups in a society. Yet it is also sometimes used as a verb to refer to the process of cohering of various groups, classes, or communities.

Finally, the strategy of reasonable accommodation is the bridge that joins public institutions and cultural diversity. It is the practice of allowing an adjustment in the practices and norms of an institution to accommodate the special needs, cultural mores, religious differences, or disability of both individuals and groups, provided the adjustment does not compromise the institution's functions and objectives, infringe on other people's rights, impose undue costs, or undermine public security or order.[42] The adjustment is the accommodation, and the conditions that it has to meet are the criteria of reasonableness. The concept of reasonable accommodation has been explicitly incorporated in employment and disability acts. For example, the US Americans with Disabilities Act of 1990 requires employers to accommodate employees with known disabilities. Similarly, human right legislation in Canada and the United States allows reasonable accommodation of cultural as well as religious practices in employment, education, and public activities.[43]

The objections to reasonable accommodation come from both the right and the left. The former view it as a compromise with public laws and social unity, and the latter regard it as a burden on minorities and a distraction from their rights.[44] Yet it has proven to be an effective strategy for inclusive policies. It is a pragmatic approach to balance diversity and common purpose.

Reasonable accommodation is an instrument of implementing multicultural rights. It allows the recognition of cultural, religious, and social differences, while preserving civic order and institutional unity. Reasonable accommodations add up over time to bring about social change.

Taken together, these concepts of reconciling the demands of diversity and unity point towards the internal heterogeneity of human societies; social and cultural homogeneity is not a reality. Multiculturalism not only makes explicit the social and cultural differences, but also recognizes them as rights. It extends the principle of equality to cultural, religious, and identity rights. Yet it does not sacrifice national solidarity. Certainly, it poses challenges and raises conflicting demands, but it remains a viable ideal for today's globalizing urban world.

Summing Up the Argument about Multiculturalism

The central idea coming out of the preceding analysis is that cultural pluralism and ethno-racial diversity are the "natural" conditions of modern, democratic, and civil-rights-based societies. Multiculturalism is the recognition of this idea. It is based on both ideological and empirical repudiation of cultural homogeneity as well as the privileging of an ethno-racial group in a nation state and society. Yet the repudiation of cultural homogeneity does not mean the absence of common purpose, institutional unity, and social solidarity. Multiculturalism does not envisage a society of socially segregated communities that have little in common. Rather, it balances the demands of a unified political and social order with the rights of constituent communities to express their identities and organize their domestic and communal lives in accordance with their cultural and religious norms and values. The following five propositions sum up the structure of multiculturalism:

1. Multiculturalism combines racial and ethnic diversity with the equality of expressing identities and enacting cultural norms and values largely in the private domain. It is an extension of the post-1960s rights revolution to culture and race.
2. The diversity of the private domain is combined with the unity of the public domain. Cultural rights of the private domain are accompanied by the appropriate inclusion of minorities' interests in the public domain.
3. Common ground is the framework of shared laws, norms, values, moral codes, etiquettes, symbols, and practices of the public sphere. It is the basis of civic order, even of a social contract that unifies all communities and groups in a society. It cultivates the sense of belonging and promotes social integration.
4. Multiculturalism is a two-sided coin, with diversity of community cultures on one side and unity of the common ground on the other. Both evolve and change in tandem.

5. The challenge in modern society is to balance diversity, equality, and unity. Almost all modes of balancing these demands, namely, assimilation, integration, and so on, end up producing a pluralistic social order. Multiculturalism is both a lived reality and an ideal of social life in times of global flows and universal rights.

These propositions will illuminate the search for patterns in multicultural cities in the chapters to come.

The diversity of people, activities, and roles has been the strength of cities. Peter Hall traces diversity as the source of creativity even in ancient and medieval cities.[45] Jane Jacobs has identified diversity as the driving force of urban economy and social life.[46] Richard Florida offers the theory that "regional economic growth is powered by creative people who prefer places that are diverse, tolerant and open."[47] Yet this diversity is sustained by the city serving as the common ground. Its collective life, shared space, services, and institutions contribute to the formation of values, beliefs, and behaviours. The bonds of citizenship promote a shared civic culture.

Cities, Citizenship, and the Right to the City

In the words of William Shakespeare, what is the city but the people?[48] More is implicit in this statement than is apparent. A city is an organized body of people of diverse backgrounds inhabiting a small area, tied together by institutions, activities, services, and technologies.

Historically, cities have always been the incubators of civilizations and the crucibles of ideas and cultures. But in our times, cities have become all the more the loci of economic growth and socio-cultural development. For the first time in human history, the majority of the world's population lives in city-regions and societies have reached the point of the "urbanization of everybody," in the apt phrase of Janet Abu-Lughod.[49]

In this era of universal urbanism, urban theories are increasingly probing the role of cities in the globalized world where ideas, technologies, trade, finance, and people move rapidly across national borders. Citizenship is viewed as the primary bond that ties together a city and lays the basis of public culture, as the uniformities of culture and nationality have eroded. Thus, questions about the rights and responsibilities of citizenship are at the forefront of urban discourse. This approach also brings out citizenship as an element of the common ground in the multiculturalism of cities.

The current ideas regarding citizenship and cities can be summed up in three themes: (1) citizenship as the basis for constructing public culture and negotiating class, race, ethnicity, and gender differences in the shared space of a city; (2) cities as the normative promise of the good, engaged, and just life, that is, as ideal types of an empathetic habitat; (3) rights to the city and diversity.

First, the ideas of citizenship and public culture are being advanced as the ties that bind cities. Nations as the sites of citizenship have history to back their claims, but transnationalism and regionalism are whittling away the role of nations, particularly in Europe. Cities are increasingly the locale of citizenship that provides services, engages citizens, and invests rights. It is the bond of citizenship that instils a sense of belonging in people of diverse identities and interests and forges a shared public culture in cities. James Holston and Arjun Appadurai observe that "place remains fundamental to the membership in society ... and cities are privileged sites for considering the current negotiation of citizenship."[50] They maintain that citizenship includes rights of participation in politics as well as civil, socio-economic, and cultural rights.[51]

Second, normative models of cities form another theme in the current literature, and are meant to address inequities based on race, class, ethnicity, and gender. Cities' long-presumed homogeneity of population and uniformity of culture have been challenged. Yet they are responsible for realizing equality of political, economic, and cultural rights across differences and for building the solidarities of a commonweal. Ash Amin's "good city" values differences, strengthens the commons, and crowds out the urbanism of exclusionary and privatized interest.[52] It is a progressive city, where people are politically aroused, engaged, and empowered. Such a city has been long dreamt about, but almost nowhere realized. Yet this has not deterred theorists from postulating it as an ideal type. Compared to this conception of the good city, Susan Fainstein's "just city" is a more pragmatic model that aims to change the system incrementally through the pressure of policies seeking justice. It envisages a city "in which investment and regulation would produce equitable outcomes rather than support those already well-off."[53] Although distributional equity is the focus of her book, Fainstein holds that social justice requires not melting away differences of race, religion, gender, and culture, but respecting them.[54] Leonie Sandercock's inclusive city is where differences are recognized and actively engaged.[55] One common feature of these models is their political philosophy, which envisages a high level of civic engagement as a way to realize equity. It puts a lot of faith in people's activism and public spiritedness.

The third theme of current urban theories expands upon the idea of the right to the city, whose articulation in contemporary discussions is attributed to Henri Lefebvre. Apart from the rights associated with civilization – the "right to work, to training and education, to health, housing leisure, to life" – Lefebvre envisages a "distinctly political space within which use-value rules and people are in the centre, participating and liberated from the segregated and fragmented social life."[56] Mark Purcell helps us understand the layered meanings of Lefebvre's conception of the right to the city, which is based on the idea of use-value as the entitlement to property in a city, leading to (1) cultivation of a web of cooperative social relations among urbanites and (2) forging a new contract of citizenship involving thoroughgoing political awakening and instilling a new consciousness.[57] The right to the city is a manifesto of an urban revolution that is both post-capitalist and post-Marxian in conception. David Harvey observes that the right to the city is not merely a right of access to what the existing system provides, but the right "to make the city different, to shape it more in accord with our heart's desire, and to remake ourselves thereby in a different image."[58] Despite these transformative goals, the right to the city has mostly been interpreted as a call to political engagement, equitable services, and the accommodation of diversity.

These urban theories shift the focus of discourse to the political sociology of cities. This shift is a reflection of the agenda of the "revolution of rights," starting from the 1960s, and the fluidity of city structures resulting from immigration and globalization. From the perspective of multiculturalism, the takeaway from these theories is that equity as an urban goal requires both accommodating differences as rights and binding different groups in ties of citizenship in a shared space. Cities are the sites for constructing such a society and citizenship requires civic engagement and political participation. It reaffirms the two-sided model of multiculturalism, namely, sub-cultural identities in private domains complemented by the common ground underlying the public domain.

Before concluding this chapter, I need to mention that the questions of the internal structure of contemporary cities and the organization of their social, economic, and cultural institutions draw on diverse bodies of theories. Even generic urban theory is tending to be city-specific, as is evident from the theoretical schools of Chicago, Los Angeles, and New

York. The ideas about specific urban institutions have to be discussed separately in the respective chapter as a guide for empirical observations. This is what I have done. It should be kept in mind that the focus of this book is on how these institutions and structures accommodate diversity and respond to the rights of differences. From that perspective, the task is to outline the critical conceptual elements defining an institution and then examine the infusion of multiculturalism into it. I will follow this approach chapter by chapter.

Conclusion

Multiculturalism is a foundational concept of this book. This chapter critically examines the concept and links it with cities and citizenship.[59] Multiculturalism is a contested terrain. The term is used with many meanings. The critical issue in multiculturalism is that of balancing the rights of diversity with the imperative of living together as an integrated and cohesive society for the commonweal. The chapter's two-sided model of multiculturalism and the idea of a common ground are its contribution to this discourse. This model will serve as the framework for analysing the infusion of a pluralistic ethos in the institutions and structures of Toronto, New York, and Los Angeles, cities that are known to be multicultural.

Chapter Three

Making Multicultural Cities

The Multicultural City as a Social Construct

A multicultural city is not just an area in which people of different ethnic backgrounds live near each other. It is a place where different races, cultures, and lifestyles are valued, their differences are recognized, and their interrelations are cultivated. The foundations of a multicultural city are laid by the ethno-racial diversity of its population.

Historically, cities always have had people of different ethnicities, cultures, religions, and languages. Yet minorities lived at the mercy of a monarch or the majority. Even in the settlers' countries of the United States, Canada, or Australia, successive waves of immigrants, such as Irish, Poles, Italians, Jews, Chinese, and Japanese, were subjected to discrimination and segregation. Only in the post-1960s era of civil and human rights, has citizenship come to mean the entitlement to social equality and the right to live by one's culture within the purview of the law and civil society. On these scores, the multicultural city is a new phenomenon.

Of course, multiculturalism alone does not bring an era of harmony and equality. Rights may exist in law, but their realization depends on the institutionalization of equality in economic, social, and cultural matters. Entrenched institutions and power politics may thwart the progress of ethnic and racial minorities. In fact, most multicultural cities have undercurrents of ethnic disparities and racial discrimination. Cultural clashes about values and practices are also not uncommon. Public controversies about the hijab (head covering) and niqab (face-veil) of Muslim women, the kirpan (ceremonial dagger) worn by Sikhs, and the profiling of black youth or targeting of Latinos for police checks agitate one city or the other. Yet rights lend recognition to people's identities and establish the rules for fairness. They ignite hopes for change.

This chapter probes the question, How does a city come to be ethnically and racially diverse? From where does its demographic diversity arise? Immigration, in-migration, and globalization are the processes that diversify the population of a city. And demographic diversification leads to social and cultural diversity. It is the arrival of people of different ethnic, religious, and linguistic backgrounds that lays the ground for multiculturalism. This process will be examined in this chapter by using the examples of the three cities under study.

Multiculturalism is also precipitated by the demands for equality of long-repressed minorities such as blacks and aboriginals. Blacks were brought in from Africa almost with the settlement of North America. The multiplicity of aboriginal cultures predates the arrival of Europeans. In the United States, despite a long history of wars and violence, 5.2 million natives made up 1.7% of the US population in 2010, including tribes made familiar by the Hollywood Westerns such as Navajo, Sioux, Cherokee, and Mohawks. Under the 1975 Indian Self-Determination and Education Act, there are 562 recognized tribal governments. Canada also has a striking diversity of aboriginal groups. It has 600 bands, adding up to the total population of 1.4 million (2011) – about 4.3% of the national population. Increasingly, aboriginals are migrating to cities and their presence is adding another patch to the cultural mosaic. But they stake their identity on being members of distinct nations. They usually keep aloof from the discourse of multiculturalism, which is primarily focused on immigrants and ethnic minorities.

The Demography of Immigration

Immigration is the primary force diversifying the racial and ethnic composition of a city and country. The United States and Canada are countries founded by immigrants who have come in waves, rising and ebbing with economic cycles, wars, peace, and the outlooks of various political ideologies. Dowell Myers identifies four waves of immigration in the United States: (1) before the 1820s Western Europeans, mainly British, came as colonizers and settlers; (2) during the period 1820–60 settlers for newly opened land came from Northern Europe, Ireland, and Germany; (3) the period 1880–1920 saw immigrants coming to work in newly industrializing cities, primarily from Southern and Eastern Europe, with a sprinkling from Japan and China in California;

Table 3.1 Percentages of immigrants in populations of the United States and Canada, 1910–2010

Year		Foreign-born as percentage of total population	
USA	Canada	USA	Canada
1910	1911	14.7	22.0
1920	1921	13.2	22.3
1930	1931	11.6	22.2
1940	1941	8.8	14.7
1950	1951	6.9	14.7
1960	1961	5.4	15.5
1970	1971	4.7	15.3
1980	1981	6.2	15.6
1990	1991	7.9	15.9
2000	2001	11.1	18.4
2010	2011	13.1	20.6

Sources: New York City, Department of City Planning, The Newest New Yorkers (2013), table 2.1, p. 10; Statistics Canada, Immigrant Status for the Population of Canada, Provinces and Territories 1911 to 2006 Census – 20% Sample Data, Catalogue no. 97–557-XWE 2006006; Statistics Canada, Immigration and Ethno-cultural Diversity In Canada, Analytical product, 99–010-X (2013).

Note: The decennial censuses in the US and Canada are held one year apart; thus, the years for the two countries' data are different.

(4) from 1970 to the present migrants, mostly from Latin America, Asia, and Africa, have been arriving under the reformed Immigration Act of 1965 to fill labour and population deficits.[1] Canadian immigration trends parallel those of the US cycles, but their ebbs and flows occur in a narrow range. Table 3.1 shows the effects of these cycles.

Immigrants are identified as foreign-born in the censuses. In the earlier wave, the percentage of foreign-born in the population peaked in 1910 in the United States and in 1921 in Canada. Then it started to decline, bottoming out in 1970 in the United States and twenty years earlier in Canada. The current wave of immigration started in 1970 and continues to rise with no end in sight. Given Canada's relatively small population (33 million in 2011), the continual inflow of 225,000 to 280,000 immigrants (temporary foreign workers apart) per year will add up to 25%–28% of its population being foreign-born by 2030.[2] With about one million legal immigrants coming per year (undocumented immigrants are in addition), the number of foreign-born in the United States in 2030 is projected to be about 14%–15% of the population.[3] Obviously, Canada is becoming a highly diversified society, with 200 self-identified ethnic groups in 2011, proportionately much more than in the United States.

The current cycle of immigration is propelled by a demographic shift that has no precedent. It is driven by a drastic drop in the birth rates of the native populations in Canada, the United States, European countries, Japan, and even China. David Foot describes this drop as the bust following the baby boom of 1947–66.[4] The total fertility rate (number of children born to a woman in her whole reproductive life) in the United States has hovered slightly above 2.0 since 1972, near the replacement level of 2.1, but in 2010 it fell to 1.9. The Canadian rate has remained in the 1.5–1.7 range, well below the replacement level, for almost twenty years.[5] With such birth rates, the natural growth of population has plummeted. Immigration is keeping up the population growth of both countries. And this is resulting in changes in the ethno-racial order. William Frey of the Brookings Institute estimates that the United States is becoming a majority-minority country, where European Whites will be a minority. This is already the case for the below-18 age group.[6]

The continuation of this trend is resulting in the aging of both nations' populations, changes in the structure of the labour force, and a realignment of the societies. But most of all it is making immigration the lifeline for both the economic growth and stability of the population. The yearly number of immigrants may rise or fall with economic and political cycles, but the need for them will remain unrelenting for decades.

Currently, immigrants are largely coming from Asia, Africa, and Latin America. Their non-European origins make their cultural and racial differences both more striking and enduring than those of the previous waves of immigrants from Northern and Southern Europe. An intergenerational chasm in ethno-racial terms is beginning to form.

The fall in birth rate is accompanied by an increase in life expectancy in both countries, resulting in increasing numbers of seniors. This greying of the population, when combined with the filling of the younger ranks with immigrants, is producing a situation wherein age differences broadly coincide with racial and ethnic differentiation. Dowell Myers identifies this situation as a call for forging a new intergenerational social contract, which binds investments in immigrants' education and integration with the security and welfare of the aging native-born.[7]

Toronto, New York, and Los Angeles
as Majority-Minority Cities: The New Reality

Cities are at the forefront of demographic change and the torch bearers of the social transformations such change brings. They are the primary destinations of immigrants. Out of the 1.1 million immigrants who came to Canada during the period 2001–6, 70% settled in three metropolitan cities, Toronto, Vancouver, and Montreal, Toronto stood out as the recipient of 40.4% of immigrants. Like Toronto, the New York and Los Angeles metropolitan areas take the lion's share of immigrants coming into the United States. From 2000 to 2009, 21% of all net immigration gains in the United States occurred in these two metropolitan areas, while they had only 10% of the national population.[8] Table 3.2 gives an overview of the population change in the three cities. It shows that the Toronto CMA population grew by a quarter in ten years, whereas New York's and Los Angeles's grew by 2%–3%. Immigrants were nearly half of the population in Toronto and Los Angeles cities and about a third in New York and Los Angeles County. They are spread out across metropolitan areas in both the central cities and suburbs. More significant from the perspective of the ethno-racial mix is that Whites had a razor-thin majority in the Toronto area and Los Angeles County, but were about one-third of the population in New York City and Los Angeles city. The trends were upward for immigrants in all three places since 2000/1 and before.

What is more revealing of the ethno-racial transformation of these areas is the ethnic and racial profile of their populations (table 3.3). Although table 3.3 presents a cross-sectional picture of the ethno-racial composition of the three cities and their metropolitan regions for the year 2010/11, it bears a story of their demographic evolution. In all three areas, Whites in general and European-origin Whites in particular were about half to one-third of the population. The Toronto area retained a thin majority of Whites, which is bound to turn into a minority with further immigration. New York and Los Angeles cities and regions had only 28%–33% non-Hispanic Whites, a drop of about 8%–13% since 1990.[9]

The steady decrease in the percentage of native Whites is the result of non-White immigrants' influx from abroad, in addition to the declining birth rates of Whites. Toronto city and the CMA region have evolved into highly multi-ethnic and multiracial places. They have almost two hundred different ethnic groups, but South Asians, Chinese, and Caribbean Blacks are the dominant groups among non-Whites. And among

Table 3.2 Demographic indicators of the three cities

Indicator	Toronto city	Toronto CMA	New York City	Los Angeles County	Los Angeles city
Population 2000/1	2,456,808	4,647,960	8,008,278	9,519,338	3,694,820
Population 2010/11	2,615,060	5,841,100	8,175,133	9,818,605	3,792,621
Change per cent	6.4	25.6	2.1	3.1	2.6
Immigrants 2000/1	1,214,630	2,032,960	2,871,032	3,011,476	1,771,965
Immigrants 2010/11*	1,252,215	2,537,405	2,971,143	3,064,140	1,781,621
Immigrants 2000/1 %	49	43.7	35.9	32	48
Immigrants 2010/11 %	48.6	46	36.3	31	46.9
Whites 2010/11 %	51	53	**44.7	**50.3	**49.8

Sources: Statistics Canada, Census of Population 2001, 2011 and National Household Survey, 2011; David Halle and Andrew Beveridge, "New York and Los Angeles: The Uncertain Future," in *New York and Los Angeles*, ed. Halle and Beveridge (New York, 2013), table 1.1, based on 2000, 2010 census data from Social Explorer (www.socialexplorer.com), pp. 5–6.

*Toronto's data on immigrants in 2011 are based on Statistics Canada, National Household Survey, Focus on Geography Series.
** Whites here refer to all Whites, not just non-Hispanic Whites.

Table 3.3 Ethno-racial profiles of the three metropolitan cities

Toronto city (2011)	Toronto CMA (2011)	New York City (2010)	Los Angeles County (2010)	Los Angeles city (2010)
All Whites 50.9%	All Whites 53.0%	All Whites 44.0%	All Whites 50.3%	All Whites 49.8%
South Asians 12.3%	South Asians 15.1%	Non-Hispanic Whites 33.3%	Non-Hispanic Whites 27.8%	Non-Hispanic Whites 28.7%
Chinese 10.8%	Chinese 9.6%	Latinos 28.6%	Latinos 47.7%	Latinos 48.5%
Blacks 8.5%	Blacks 7.2%	Blacks 25.6%	Blacks 25.6%	Blacks 9.7%
Filipinos 5.5%	Filipinos 4.2%	Asians 12.7%	Asians 13.9%	Asians 11.4%

Sources: As in table 3.2.
Note: Columns do not add up to 100% because the categories are not mutually exclusive. For example, "All Whites" include non-Hispanic Whites.

Whites, Italian, Jews, Russians, and Portuguese are in sizeable numbers. Furthermore, such categories as Chinese, South Asia, and Blacks include a wide range of ethnic and nationality groups, for example, Chinese from Hong Kong, the mainland, and Southeast Asia; Indians, Pakistanis, Tamils, and Bangladeshis among South Asians; and Blacks from Jamaica, Trinidad, Ghana, and Nigeria. Again, this diversity is almost entirely the development of post-1965 immigration. The Toronto CMA has suburban towns and cities, where visible minorities (non-Whites) formed the majorities in 2011, such as Markham (72.35%) and Brampton (66%).

The demographic composition of New York and Los Angeles cities parallels that of the Toronto area in the percentage of immigrants, but they have a distinct thrust towards Latinization. New York City had about 29% Latinos/Hispanics, who edged out Blacks (27%) as the non-White minority. In the Los Angeles area, Latinos/Hispanics were about 48% of the population (table 3.3). They came from many countries, such as Mexico, El Salvador, and Guatemala, and are at the cusp of becoming a majority. Asians are another category of immigrants who are increasing in numbers and cultural influence. There are cities in Los Angeles County, such Monterey Park, Walnut, and San Gabriel, where Asians were about 60% of the population, Chinese alone being 33%–40%, in 2010. American Blacks are, like Whites, losing proportionately to immigrants. Their share of the population in the Los Angeles area was in the 9%–10% range. In all areas their share has decreased by 3%–4% since 1990.[10]

These cities are turning into majority-minority places, where no single ethno-racial group is in the majority (50% or more), but minorities (including immigrants) together form the majority. All the more significant from the point of the future demographic composition of these cities is the fact that the percentages of Whites among children and youth five to nineteen years old were only 43% in the New York–New Jersey metropolitan area and 17.3% in Los Angeles County in 2013.[11] These cities are going to have non-White majorities of diverse ethnic and racial backgrounds.

In the United States also, majority-minority cities are beginning to be the norm. In the top 100 metropolitan areas, 58 primary cities were majority-minority places in 2010.[12] Furthermore, minorities are not only concentrated in central cities, but also have spread out to the suburbs.[13]

With people of different races and cultures living in a city, it is inevitable that a variety of cultural communities will be formed. Immigration breeds multiculturalism, which is then nurtured by civil rights and reinforced by globalization. New citizens of Russian, Hispanic, or Chinese origins, for example, imprint a city with their food, music, and institutions, forming distinct communities and creating economic niches. Their presence diversifies both the local culture and economy.

Immigration as a Force for Social Change

People have been migrating from one part of the world to another since time immemorial. There is no country whose population is pure-bred native. Every population group has evolved by absorbing successive waves of immigrants. Europe, which is regarded as the region of homogeneous populations, was historically a land of migrants. Celts, Romans, Saxons, Danes, Normans, and Huguenots, for example, successively swept across Britain, and today's English are their mixed progeny. India has been swept at various times by Aryans, Mongols, Afghans, Mughuls, and Persians. Migration has been a continual part of population growth in human history. The current misgivings about immigration arise from the ideology of nationalism and the notion of the singularity of national identity.

Present-day immigration in North America differs from its precedents in many ways. It is ethno-racially more diverse. It brings immigrants of varying educational and professional attainments. In the early twentieth century, immigrants were predominantly peasants and industrial workers starting at the bottom of the social structure. Among present-day immigrants is a steady stream of highly qualified professionals who settle in as professors, doctors, corporate executives, and investors, though many similarly qualified people experience difficulties and bias in getting their credentials accepted. The result is an injection of social diversity in all strata of the host societies.

There is a noticeable "brain gain" in the receiving societies, particularly in the United States, Canada, Australia, and Britain. About 52% of the start-ups in Silicon Valley, California, were established by immigrants,[14] including companies such as Intel, Sun Microsystems, and Google. In 2006, foreign students, both permanent and temporary residents, earned between 46% and 61% of doctorate degrees in sciences, mathematics, and computer sciences.[15] In Britain, over half of the registered doctors from 1992 to 2002 were educated abroad.[16] The situation in Canada is not much different. By 2000, it had attracted 2.7 million foreign graduates. These highly qualified and entrepreneurial immigrants have contributed to the economic growth of all these countries.

Immigration not only affects the composition of population in the present, but also transforms the ethno-racial structure of a society for the future. The children of immigrants, second and third generations, grow up like the native-born, imbibing societal culture and national identity, but their "roots" remain embedded in ethno-racial identities. Cumulatively, immigration builds up ethno-racial communities and sets in motion processes that restructure the ethnic and racial basis of a society.

Another process that reinforces the diversification of urban cultures is globalization, namely, the rapid movement of capital, labour, ideas, and products across nations and the resulting interlinking of economies and societies. Globalization circulates business executives, investors, professionals, workers, and tourists through cities, fostering diversity of lifestyles and eroding cultural homogeneity. It reinforces the effects of immigration by facilitating the movement of people and products.

How immigrants fare depends on the policies and social conditions of the recipient country as well as of the local economy and social ethos in specific cities. Does a country have pervasive discrimination, with ethnic or racial prejudices; does it offer equal opportunities to immigrants? These are some of the determinants of the integration of immigrants. Similarly, a city's labour market, political system, immigration history, housing stock, connections with the home society, and so, on influence the integration of immigrants.[17] Still, immigration evokes strong emotions, both positive and negative, and is a topic of fierce public debates.

The Economic Discourse on Immigration

Taking the economic arguments first, political views about the social costs and benefits of immigration have, strikingly, changed very little over the past two centuries, if not longer. During the nineteenth and early twentieth centuries in the United States, immigration was actively sought by businessmen, industrialists, and farmers, who wanted a plentiful supply of workers to keep wages low and businesses booming, whereas labour agitated against unrestricted immigration, complaining that newcomers take away jobs and depress wages. The same arguments by similar groups reverberate in contemporary debates about immigration. Current arguments for and against immigration can be summed up in the following themes:

Pro-immigrant arguments
- Immigrants increase the supply of labour and provide the workforce for economic growth.
- Natives move into managerial and professional occupations and the jobs that they do not want are filled by immigrants.
- Immigrants do not affect the wages of native groups to any appreciable degree, but form a segmented labour market.
- Immigrants expand the demand for goods and services and stimulate economic growth.
- Immigrants pay taxes and increase public revenues.
- Immigrants bring new entrepreneurial energy and creativity.
- Immigrants are young and their contributions sustain the social security system for the retiring natives.
- Immigrants revive the deteriorating housing stock and declining cities.

Arguments against immigration
- Immigrants take up jobs and displace native workers, particularly of minority communities such as Blacks in the United States or Aboriginals in Canada and Australia.
- Immigrants swell the labour force and depress wages.
- Immigrants increase the public costs of health, educational, and welfare services, and become a public burden.
- Immigrants increase social disparities between the rich and poor as initially they add to the ranks of low-income households.
- Immigrants settle in cities, changing their social structure and cultural organization, while small towns and rural areas continue to be the strongholds of historic communities. These sociocultural differences frame national politics.
- Local governments bear the costs of immigrants' settlement, while national governments reap the economic and financial benefits, which leads to a fiscal crisis for cities.
- To these arguments, the current notions of environmental sustainability have added a new theme of controlling population, even advocating zero growth and a resulting restriction of immigration.

These opposing propositions permeate the literature about the social and economic impacts of immigration.[18] Philippe Legrain's book *Immigrants: Your Country Needs Them* is outspoken in its advocacy for immigration; it builds its case through a sweeping examination of the pro and con arguments.[19] Doug Saunders argues passionately about the energy and enterprise that immigrants bring to cities all over the world.[20] On balance, the weight of evidence, both historical and current, shows that immigration contributes to the economic development of host societies, while its slightly negative impacts on unskilled workers and minorities are minor and transitory. Yet, like all good things, immigration also has to be in the right amount. It can be too large and burdensome, particularly in recessionary times or under the conditions of resource constraint if there is a mismatch between the immigrants' skills and the host society's demand for labour.

Although the economic arguments tend to favour immigration, public opinion in most receiving countries tends to be leery of immigrants.[21] There is often an undercurrent of scepticism about immigrants' adaptability and loyalty to the host society. The public perceptions about immigration are determined by social and cultural concerns.

Immigration, Social Life, and Culture

Immigration injects persons of different languages, beliefs, and ways of living into the resident population of a city and society. Often immigrants are young and energetic. They realign the social structure of communities where they settle, changing its ethno-racial mix, swelling the ranks of the lower strata, while investors, executives, and professionals among them find footholds in the middle and upper strata. But it is the ethnicity and race of immigrants that have the transformative impact on the social organization and culture of a city.

Immigrants' path of inclusion in a society has differed by their race and ethnicity. Northern Europeans migrating to the United States, Canada, and Australia, or within their own region from one country to another, are readily absorbed. Southern and Eastern Europeans have had more difficulty, while Africans, Caribbeans, or Asians continue to be regarded as foreigners for generations. For example, Japanese and Chinese born in Canada or the United States whose grandparents came in the early twentieth century are often asked, "When did you come?" or "Where are you from?" Thus, immigration draws a fissure in the social structure along racial-national lines, creating what John Porter, a distinguished Canadian sociologist, called a vertical mosaic.

Immigrants bring with them different ways of life, religions, and languages, giving rise to new communities and institutions. Immigrants of similar origins bond together for material and emotional support and form subcultural communities. They organize religious institutions, mutual-aid associations, language schools, ethnic clubs, literary events, fairs and festivals, and so on, primarily to meet their cultural needs, but in the process diversifying the institutional base of a city. These institutions sustain the sense of community and identity among the second and third generations. Many of these social organizations find concrete expression in the form of ethnic enclaves and business districts, resulting in the restructuring of a city.

It may be pointed out that the diversity of beliefs and behaviours is not just a matter of individual differences, but a manifestation of variations among groups. It is not only that a Jewish or Chinese person lives by her values and norms, but also that Jews and Chinese form communities and associations. This is what lays the bases for multiculturalism.

Large-scale immigration reorders the everyday life of a city. On the street, one encounters different languages being spoken and the national language in unfamiliar accents. New foods of exotic tastes and smells are added to the local gastronomic repertoire. Costumes once seen in movies and on television are now found to be the attire of one's neighbours. One witnesses different ways of treating children, new forms of families, and unfamiliar customs. A pluralistic culture comes into being.

Sociocultural Concerns about Immigration

In the host societies, there has always been a current of scepticism and apprehension about immigration. In economically good times, this current is submerged by the need for immigrants' labour and by plentiful economic opportunities. In bad times, the anti-immigration sentiments surface and concerns about "foreigners" taking away jobs arise. In recent times, civil-rights legislation and the awareness of racism inhibit open venting of xenophobic feelings. The public discourse about immigration tends to be couched in terms of its social costs and benefits and often not in nationalist ideologies. Yet there is a remarkable historical continuity about the public concerns with immigration.

For example, in Canada, a proclaimed multicultural country, the early-twentieth-century immigrants were described by J.S. Woodsworth (founder of the socialist Cooperative Commonwealth Federation, or CCF) as "strangers within our gates." He particularly singled out the "Orientals" (Japanese, Chinese, and Hindus), who "cannot be assimilated" because they have their own "virtues and vices."[22]

US president Teddy Roosevelt (1904–8) in a speech in 1915, said, "There is no room in this country for hyphenated Americanism ...

We must unsparingly condemn any man who holds any other allegiance."[23] These sentiments resonate even today, particularly among those opposed to substantial immigration, even though the idiom of discourse has changed.

Samuel Huntington, the Harvard academic famous for the "Clash of Civilizations" thesis, suggests, "Mexican immigration is heading towards the demographic *reconquista* of areas America took from Mexico by force in the 1830s and 1840s."[24] The Federation for American Immigration Reform (FAIR), a lobby group for restricting immigration argues that present-day immigrants neither make significant contribution to the economy nor are they willing to adapt to the American culture.

After the terrorist attack on the World Trade Center in New York on September 11, 2001, immigration has been linked with national security. Immigrants, particularly from Muslim countries, are viewed with suspicion, and have the additional burden having to prove their loyalty to Western societies. Even in Canada, a country largely unaffected by terrorism, there is a current of disquiet about immigration and national security.

In Canada, despite its proclaimed policy of multiculturalism, the cultural adaptation of immigrants remains a simmering concern. The conservative, but influential, Fraser Institute puts forward for national debate questions of immigration policy. A theme in the Fraser Forum is that the growing presence of immigrants from non-European countries places demands on Western democracies to accommodate the non-Western cultural values of immigrants.[25] Janice Stein, the director of the Munk School of Global Affairs at the University of Toronto, worries about immigrants' religious and cultural practices that conflict with Canadian values of gender equality and freedom of belief.[26] This theme is expanded to include many practices of immigrants that purportedly clash with liberal values. Apart from gender inequality, fears are raised in the media about acts such as forced marriages and honour killings and the Sharia law being brought into Canada and the United States. Thus, immigrants' social and cultural adaptations, or lack thereof, are matters of public concern.

The themes regarding immigrants that commonly reverberate in the anti-immigration arguments can be summed up thus: (1) a failure to integrate; (2) divided loyalty; (3) not sharing the responsibilities of citizenship; (4) a poor sense of national belonging; and (5) non-committal to such basic national values as democracy, freedom of expression, and tolerance. All these objections have been largely unproven, at least in countries such as Canada, the United States, and the United Kingdom.

They arise from the obvious "otherness" of immigrants. There is no evidence that immigrants have been disloyal while holding on to their ethnic or national identities, either in the past or in the present. The interesting part here is that in this globalizing world, many native-born Canadians, Americans, and Europeans live abroad, marry, and hold dual citizenship. Notions of national loyalty are undergoing rapid change. Even the idea of nationality is being stretched to accommodate dual nationality and transnationalism.

Of course, there are many voices supporting immigration and the cultural diversity that comes with it. Michael Adams, one of Canada's leading pollsters, in his book *Unlikely Utopia*, knocks down one-by-one points raised to argue that immigrants have little incentive to integrate into Canadian society owing to its multiculturalism policies. He finds, supported by evidence, that immigrants follow Canadian laws and adopt a Canadian way of life while retaining their identity. He observes that overwhelmingly (75%) Canadians believe that immigration has a positive influence on Canadian society.[27] Joseph Berger, a *New York Times* columnist, sums up his observations of New York's ethnic neighbourhoods and the transformation of communities all across the United States by immigrants with an allegory: "They are merging – conspicuously and inconspicuously – into the mighty American river. It is a very different river from one that flowed a half century ago, fed by streams and rivulets we knew little about only yesterday, but it still rolls mightily and charmingly along."[28] Similarly a majority of academics in both the United States and Canada, two neighbours of contrasting models of immigrants' absorption, that is, the melting pot versus the mosaic, conclude that immigrants both transform and strengthen national cultures and values.[29]

Living Here and There: Transnational Immigration

There is another lens through which immigration is beginning to be viewed. The historic image of immigration is that of a population stock moving from one country to another, making the latter their only home. Migration now is increasingly taking the form of a flow, or rather circulation, of people between two or even more countries. One lives both here and there. Immigrants go back and forth, send money to relatives, build homes and vacation in their homeland, invest in businesses, send children to learn "our" culture and keep up with the homeland news on a daily basis. Native-born Americans and Canadians work and live in many countries, linking cultures and building cosmopolitan lifestyles.

A global economy, jet travel, inexpensive international calls, the Internet, Facebook, and Skype has scrambled the boundaries between cultures. Kwame Appiah argues that "contamination" of cultures, not the ideal of purity, has been the order of human societies.[30]

Diaspora is a way of living in two cultures. It is now a common experience, not limited to the jetsetters of what Pico Iyer calls the Global Soul.[31] It is not just a transitory phase in the assimilation of immigrants. Transnationalism is increasingly the outcome of immigration. Michael Valpy, a Canadian newspaper columnist, says that "we are becoming a land of global citizens."[32]

How immigrants live in two countries is highlighted by a city councillor of Hackensack, New Jersey, who was an elected official in the United States while also running for the Senate in Colombia. Mexico and Italy have set aside seats in their legislative bodies for representatives of their diasporic communities. Non-resident Indians (NRI) have been the main drivers of the software industry in India. The European Union is now a community of transnationals. Citizens of its member countries can live, work, retire, and travel anywhere within its boundaries without any restrictions. Millions of workers in international agencies and transnational corporations spend years abroad as sojourners. Temporary workers admitted into the United States, Canada, the Gulf countries, and Europe form a sizeable percentage of the resident populations. Their presence is nothing but a form of transnationalism. Approximately 3 million Canadian citizens live abroad.[33]

The United States, Canada, Britain, Italy, Mexico, the Philippines, and many other countries now allow its citizens multiple nationalities. It is estimated that 15% of the population of Western countries have more than one citizenship.[34]

Transnationalism is realigning the process of immigrants' integration. No matter how much conservatives decry multiculturalism, the erosion of cultural and political boundaries is a process independent of immigration. You cannot have a global economy, free trade, and transnational corporations and not have cultural diversity. Immigration builds on these trends.

Conclusion

Canada and the United States, as well as most other Western countries, are dependent on immigration to avoid the shrinking of their populations. Their natives' birth rates are below replacement levels and continual immigration for years to come is their only hope for fuelling economic growth, replacing an aging workforce, and maintaining social stability. Immigration now occurs in the sociopolitical environment of charters of rights, freedoms of expression and religion, as well as civil equality. Immigrants come into countries where globalization is celebrated, opportunities for trade and employment abroad are actively pursued, and multinational corporations are courted. This environment nurtures cultural pluralism and promotes diversity of beliefs and practices, which are further helped by instant access to the global media. Thus, immigration builds the human base for multiculturalism. Cities are the places where multiculturalism thrives. Canada embraces multiculturalism officially. The United States practises it without proclaiming multiculturalism as its creed.

Both immigration and its associated multiculturalism are not without problems. There are challenges of acculturating immigrants to national values and institutions. There are misgivings about immigrants' integration and apprehension about their cultural adaptation. Yet there is progress, largely through the construction and reconstruction of the common ground that is necessary for the functioning of a national society and state. The processes of balancing cultural diversity in some areas of life with building a common ground in others are the indicators of multiculturalism that subsequent chapters will examine.

Chapter Four

The Social Geography of Multicultural Cities

Socio-spatial Organization

The most talked-about expression of a city's multiculturalism is its offerings of ethnic restaurants and eateries. Beyond gastronomic multiculturalism, there are many other expressions of cultural pluralism imprinted on the urban landscape. Among them are ethnic enclaves and business districts, religious and community institutions, as well as architectural idioms. These are the footprints of different groups on the city space or, in other words, the spatial organization of ethno-racial communities. Spatial organization represents the social geography that links together "the places, the people and the practices."[1]

In this chapter, I will analyse the geographical patterns of the distribution and organization of ethno-racial groups' activities and institutions. It will give an account of how multiculturalism permeates the urban structure. The questions addressed in this chapter are: (1) how are ethno-racial groups organized residentially, commercially, and institutionally in multicultural cities as observed in Toronto, New York, and Los Angeles?; (2) how are ethno-racial identities reflected in urban space and what processes bring that about?; and (3) how does the city act as an instrument of integration?

The Internal Structure of the City: Theories and Models

People, activities, and their linkages are distributed in the space of a city in identifiable patterns. These patterns form the internal structure of a city. Underlying these patterns are social, economic, and spatial processes that sort out activities and people by location, forming districts or areas of distinct functions and socio-economic identities. Urban theorists have formulated models to explain both a city's internal structure and the processes that bring it about. To understand how ethno-racial differences affect the internal structure and inject pluralism into it, I will begin with a brief review of those theories and models.

The most common image of a city is that of a circular built-up area, divided into residential zones of decreasing densities from the centre to the periphery. At the centre of this built-up area is a high-density district of commercial-business activities. This image owes its origin to the Chicago school's Ernest Burgess, whose concentric zone hypothesis (1923) conceived the city as a product of ecological processes.[2] Other classical models include (1) Homer Hoyt's sector theory (1939), envisioning the city as made up of pie-shaped sectors differentiated by high, medium, and low rental housing, which corresponded to the residences of the associated social classes; and (2) Chauncy Harris and Edward Ullman's multiple-nuclei hypothesis (1945), which postulates the city as structured around multiple centres that serve as the anchors of its development and expansion.[3]

From these models the analytical parameters of urban structure can be deduced, that is, functions (residential, commercial, industrial), socio-economic characteristics of residents, density of development, locations of activities, and interrelations. Even recent theories of urban structure use these variables as the basis of their models. The technological, economic, and social changes of the late twentieth century have made the urban structure malleable. Cities have spread out, suburbs have grown into veritable cities, shopping malls have realigned the commercial order, and the electronic revolution has drastically diminished the resistance of distance. These changes have realigned the urban structure and given rise to new urban theories.

The Los Angeles school of urbanism projects Los Angeles as the model of a post-modern city, lacking a strong centre. It views the city to be cellular in structure, divided into autonomous places by function, culture, and location. It envisages the growth impulse to work from the outside to the central core, reversing the conventional view. The city is fragmented into functional-sociocultural districts, such as edge cities, ethnoburbs, theme parks, gated communities, corporate citadels, and command and control centres.[4] Such a city has many cores and is held together by political institutions and infrastructure. This West Coast view of the city is contrasted with what has been called the New York school, wherein the centre (e.g., Manhattan) holds strong, linking together different classes, ethno-racial groups, and activities, the growth impulses radiate out, and city life has an edge over suburban living.[5] The New York model does not envisage concentric zones, but visualizes a strong central city complemented by suburban communities of distinct identities and politics.

These models point to the fact that the structuring of cities is affected by their history, economy, politics, and social-functional mix. Such factors interact with the conventional parameters to produce new forms of urban structures, which fall along a continuum of structural models ranging from centred to decentred, nucleated to formed by a network of functional districts. These differences lend some uniqueness to a city, even in this age of homogenizing globalization. Ethno-racial diversity permeates these structures, creating distinct patterns of geographic distribution of racial and cultural groups, their institutions, and their symbols. Yet, the urban structure serves as the ground for the expression of ethno-racial identities.

The Geography of Multiculturalism

The embedding of ethno-racial diversity in urban structure takes the form of clustering by racial and/or ethnic identities, as well as their diffusion among existing institutions. The clustering of ethno-racial groups leads to the formation of ethnic enclaves, neighbourhoods of diversity, and commercial centres of distinct cultural flavours. Ethno-racial identities are implanted separately as religious and community institutions, signs and symbols in different languages, and customs and festivals of distant lands in the urban landscape. These are the manifestations of multiculturalism in the urban space.

Talk about a multicultural city and the conversation veers into discussions of Chinatown, the Latin barrio, or an Italian district. These are ethnic enclaves where not only ethnic foods, exotic dress, and strains of music from distant lands can be savoured, but also where ethnic commercial establishments flourish, and the distinctness of community life can be experienced.

Contemporary ethnic enclaves have sprung up in suburbs and even beyond. They restructure the metropolitan space, breaking the monotonous homogeneity of traditional suburbs. Witness the vibrancy of ethnoburbs of Sikhs in Brampton (Toronto Metropolitan Area), Chinese in Monterey Park (Los Angeles County), or Russians in Brighton Beach (New York), and even in second-tier cities such as the Somali neighbourhood in Minneapolis. I will explore in detail the role of ethnic enclaves in the three cities later. For now some other elements of a multicultural city's spatial organization need to be pointed out.

Ethnic malls, community centres, and places of worship (synagogues, mosques, temples, or Eastern Christian churches) are some of the other institutions reflecting cultural diversity. They create new focal points around which distinct rhythms of activities and travel patterns evolve. For example, mosques have large weekly congregations on Friday noon in addition to the daily five prayers; the Jewish Sabbath is observed Friday–Saturday, resulting in a different temporal cycle for some parts of a city. Also ethnic malls, commercial centres of a particular ethnic theme such as Chinese or Korean, become more than shopping places. They become tourists' destinations and communities' focal points. One by one these elements add up to realign the cultural landscape of a city, and reorganize its functional hierarchy.

The clustering together of similar people and complementary activities is a basic process of city development. In a multicultural city, the clustering also takes place along ethnic and cultural lines, in addition to those of class, activity, and lifestyle. Bill Bishop's book *The Big Sort: Why the Clustering of Like-Minded America Is Tearing Us Apart* discusses the recent trend in American cities towards the formation of neighbourhoods where people of similar interests, beliefs, and lifestyles congregate.[6] The urban property market is the primary mechanism of social sorting and clustering.

Ethnic Enclaves

An ethnic enclave is usually a residential cluster of a particular ethnic group that has built a community life by developing corresponding institutions, businesses, and services. It is more than the residential concentration of an ethnic group. The term enclave is also used to describe networks or clusters of ethnic businesses complementing each other in activities and resources. Alejandro Portes and Robert Bach initially used the term Cuban enclave to describe Cuban immigrants' economy in Miami, Florida.[7]

Enclaves are "voluntary" clusters in the sense that they emerge from individuals' choices. Their residents have mostly chosen to live there. These choices are mediated through the housing and real estate markets. While the structural conditions of a city's market do frame the choices, there is no guiding hand or systemic discrimination that herds an ethnic group into a particular area. In this respect, enclaves normally represent the promise of modern multiculturalism, namely, ethnic minorities exercising their civil and human rights to live where they can afford to and feel comfortable.

Enclaves, Ghettos, and Citadels

The history of racial discrimination leading to the ghettoization of racial minorities haunts the discussions on enclaves. Aren't enclaves just ghettos of immigrant minorities? It is a question often raised in public discussions. Newspapers and television often paint enclaves as breeding grounds for poverty, exclusion, and even crime.[8] Yet the same media in their reporting of a city's places of colour and excitement laud ethnic enclaves.

Scholars make distinctions between enclaves and ghettos. Peter Marcuse defines a ghetto as an area of spatial concentration created by the dominant society to separate and limit a perceived inferior racial or ethnic group.[9] A ghetto is a product of exclusion and poverty. It is also typically an area of physical and social blight. By these criteria, most contemporary urban enclaves are not ghettos. They are vibrant and are found in both the old and new parts of cities. Observers from both Europe and the North America affirm this point of view.[10]

The rich and powerful have often congregated together, creating an area of concentration to preserve their exclusivity and maintain their distance from lower classes. These enclaves of the rich and powerful, increasingly surrounded by walls and security gates, have been called citadels.[11] Contemporary ethnic enclaves are neither ghettos nor citadels, but communities of a distinct identity.

Of course, there were ethnic neighbourhoods in North American cities even in the early twentieth century, another period of high immigration. These earlier immigrants lived in poor neighbourhoods of rooming houses and walk-up apartments. The Jewish neighbourhood of the Lower East Side in New York was complemented by Polish, Irish, German, and Italian areas, along with small concentrations of Greeks, Armenians, and Chinese. These neighbourhoods could be described as pre-multiculturalism enclaves; their social life certainly had the institutional bases of enclaves. Yet they were largely the only places open to immigrants. They were tolerated, but barely so.

Toronto has a parallel history. Despite being a city largely peopled by immigrants, the British heritage was taken to be the norm and Eastern and Southern Europeans who came in the period 1910–30 were viewed as foreigners. They settled on the fringes of the downtown, near industrial establishments, in blighted neighbourhoods. Kensington Market, the Junction, and Cabbagetown are some of the neighbourhoods in the ethnic chronicle of the city. They were called "foreign colonies" and their residents were regarded as a "menace in our civilization unless [they] learn to assimilate the moral and religious ideals and standards of citizenship."[12]

Los Angeles began as a Mexican village, dating from before the United States' conquest of California in 1848. The city's incorporation in 1850 was followed by waves of migrants from the US Midwest and the South. A broad territorial distinction by ethnicity emerged, with Mexicans on the east side, Whites on the west side, and Blacks in the south. Within these broad sectoral divisions, there were fine-grain concentrations of Jews, Armenians, Chinese, Japanese, and Blacks. The city and region are now organized as neighbourhoods, many with a distinct ambience, surrounded by crisscrossing highways. It is a multipolar city whose historic geographic divisions have been changing, with new suburban ethnic enclaves appearing far out in Los Angeles County.

Historic ethnic neighbourhoods differ markedly from contemporary enclaves because of the social and cultural rights of minorities and the shifts in political ideology and the moral order. The new enclaves may be viewed with apprehension of their impeding social integration in some quarters, but they exist as the expressions of the rights of those who chose to live there.

The Benefits and Costs of Enclaves

By concentrating in an area, people of a particular ethnic background form the critical mass necessary for the viability of businesses and services catering to them as well as for political influence. The demographic concentration creates economies of scale for businesses and services and lays the ground for associations, clubs, religious institutions, fairs, and festivals. Schools can offer "English as a Second Language" classes, doctors and dentists who speak ethnic languages and are sensitive to the cultural mores of residents can establish practices, libraries can offer books and videos in home languages, and public programs can be tailored to the needs of local clients. Even local governments and other public agencies can economically deliver social services and citizenship programs. In sum, a wholesome community life in culturally sensitive ways becomes possible. These are the benefits of enclaves.

For a city, enclaves provide new points of interest and their commercial areas in particular become centres of attraction for tourists. They inject new architectural and aesthetic idioms into the urban landscape.

There are social costs associated with enclaves, however. Depending on how exclusive an enclave is, it can isolate an ethnic group, especially immigrants, and slow down their integration. In particular, concerns are raised about schools becoming ethnically or racially segregated, thereby denying children the experience of relating with others. Yet this condition is endemic to schools because neighbourhoods have concentrations by class if not by race or ethnicity.

Most of the objections to enclaves arise from their potential to impede the social mixing of ethnic groups.[13] This argument has to be examined in the context of modern urban life. Neighbourhoods are a minor site for social interactions. In contemporary urban living, neighbourliness is a social relation of politeness and shared interest in the safety and welfare of an area. Normally there are relatively limited interactions among adults in a neighbourhood, except in emergencies and contingent purposefulness. In a city, it is highly unlikely that residents of enclaves will not encounter others in their daily life. Schools, workplaces, parks, stadiums, streets, and shopping malls are the sites of interactions in cities. Enclaves do not affect these sites of integration.

Furthermore, enclaves have internal diversity. A Chinese enclave, for example, could have people from the mainland, Hong Kong, Taiwan, or Indonesia who differ from each other in language, history, and identity. Enclaves also change over time. Some residents move out, others move in with changes in job and the family life cycle. An enclave that has a high concentration of one ethnic group today could become "mixed" tomorrow.

At this juncture, an empirical examination of ethnic enclaves in Toronto, New York, and Los Angeles is in order. It will clarify how enclaves are formed and what role they have in creating the geography of multiculturalism.

Ethnic Enclaves in the Toronto Metropolitan Area

The Toronto CMA is the metropolitan area centred around the city of Toronto. It has nineteen municipalities in an area of 5903 square kilometres. It functions as an integrated economic and social region. Its population of 5.82 million (2011) grew by about 100,000 persons per year in the

2000s, mostly by immigration. As described in chapter 3, immigrants are transforming the city-region into a majority-minority area.

The Toronto city-region fans out from the shore of Lake Ontario. The city's downtown has been the pivot of the urban area, but the centres of suburban municipalities serve as subsidiary focal points, turning the whole region into a multi-nuclei area. There are clusters of high-rise apartments sprinkled over vast tracts of residential subdivisions in suburbs. Many of these residential clusters have turned into ethnic enclaves with the development of ethnic malls, places of worship, and community institutions.

A recent study of the Toronto CMA's ethnic enclaves maps the concentration of six major ethnic groups by census tract (CT), which is an area of about 4000–6000 residents.[14] The groups whose residential concentration has been plotted are Jews, Italians, South Asians, Chinese, Portuguese, and Caribbeans.

The study mapped two levels of concentration of these groups.[15] Primary concentration occurs in a CT when a particular ethnic group is the majority, namely, 50% or more of the population. A secondary concentration occurs if an ethnic group does not form the majority but the single largest group, often between 25 and 49% of a CT's population. Map 4.1 combines contiguous secondary and primary concentrations to highlight the areas of concentration of the ethno-racial groups, which had some degree of residential concentration. It was found that these residential clusters were enclaves, as they had the corresponding ethnic places of worship, malls and stores, services, and associations.

What map 4.1 shows is that five of the ethnic groups have carved out areas of distinct identities. Three of those groups are historic European immigrants – Jews, Italians, and Portuguese – whose second and third generations have grown up in Canada. Yet their ethnic identity remains strong. A Jewish enclave aligned along Bathurst Street, shifting out towards northern suburbs, forms the core of Toronto's west end. The Italian enclave radiates northward, from the city's historic Little Italy along College Street. Its residential centre of gravity is now in Woodbridge, a north-western suburb, though its commercial heart and many institutions remain in the city of Toronto. The Portuguese enclave is entirely at the core of the western half of the city, between Dundas and College Streets. These three enclaves are changing internally, but they are not expanding, as there is little immigration to feed their expansion.

The story of enclaves currently is the tale of immigrants becoming ethnics, that is, making a place for themselves in civic life. As the new immigrants are from Asia, Africa, and Latin America, the formation of

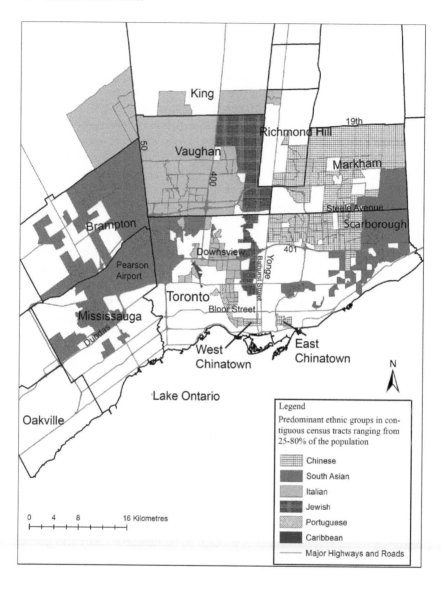

Map 4.1 Toronto's ethnic enclaves, 2006 (adapted from Qadeer, Agrawal, and Lovell, "Evolution of Ethnic Enclaves")

Photo 4.1 Mexican food in Korea Town, Toronto (courtesy Douglas Hildebrand)

their enclaves is an indication of their settlement and an expression of contemporary multiculturalism.

The Chinese have a historic footing in the centre of the city of Toronto with two Chinatowns, the western one on Spadina Avenue and the eastern on Gerrard Street. They are the markers of historic Chinese migrations of the mid-twentieth century. The new waves of Chinese immigrants have vaulted into the suburbs, forming a large enclave in the north-eastern quadrant of the metropolis, in Agincourt, Markham, and Richmond Hill. Some of the biggest Chinese malls are in this area. This enclave has grown with both an infiltration-displacement cycle in the housing market and by the building of new dwellings.

South Asians are another group that have formed two large distinct ethnic enclaves, though these are spread out and diffused. One is on the eastern periphery of the metropolis and the other is in the north-western quadrant, extending from Rexdale to Brampton. Their places of worship, strip malls, and service establishments anchor these enclaves.

African Blacks and Caribbeans do not have an enclave. Only two isolated CTs with secondary concentrations had emerged by 2006. A historic racial divide has not influenced Toronto's ethnic geography.

A look at map 4.1 shows small enclaves in the city of Toronto but an arch of enclaves in the suburbs forming a semicircle from east to west moving northward along a broad front. This is the new structure of the metropolis.

The suburban locus of enclaves is indicative of the new housing stock and high percentage of homeownership. It underlines the difference between contemporary enclaves and the old ethnic neighbourhoods. Another feature of the Toronto area's enclaves is that though ethnics are the dominant group in their respective enclaves, they are not exclusive places. The study from which map 4.1 has been taken concludes that all five groups were minorities (ranging from 35% to 45%) in their respective enclaves, though in the small cells of the cores they could be the majorities.[16]

The thriving enclaves of even long-settled groups, Jews, Italian, and Portuguese, show that social integration does not necessarily lead to spatial assimilation. Ethnicity remains a defining element of people's identity. It realigns the spatial structure of a city. As Ceri Peach sums up, the logic of spatial assimilation that the "social melting pot also melts the spatial enclave" does not happen.[17]

These enclaves are enduring, as a follow-up study by the University of Alberta's Sandeep Agrawal shows. He has mapped Census Canada's National Household Survey data for 2011 by CTs. His yet unpublished study shows that between 2006 and 2011, new ethno-racial groups, namely, Filipinos and Caribbeans/Blacks have also formed residential concentrations. Whereas the Jewish enclave has thinned out, with the shifting of many concentrations from primary to secondary, Chinese as well as South Asian enclaves have expanded a bit. Yet the overall pattern has remained as in 2006, except for some shifting of boundaries on the margins. The point is that with continuing immigration and the development of ethnic stores, places of worship, professional practices, and other institutions, enclaves become engraved on multicultural cities' geography.

The following points sum up the structure of ethnic enclaves in the Toronto metropolitan area, as observed in successive studies.[18]

1. Long-established ethnics of many generations, such as Jews, Italians, and Portuguese, have enclaves as thriving as the recent immigrant groups. Most large immigrant groups of distinct culture/religion form some residential concentrations, which turn into enclaves.
2. Ethnic enclaves have restructured the residential and commercial geography of the metropolitan area. They form an arc in suburbs around the city of Toronto.

3. Ethnic enclaves are dominated by particular ethnic groups, who are proportionately the largest group but are numerical minorities (35%–45%), except that some census tracts in their cores have more than 50% population of the dominant group. They are not the exclusive group in the neighbourhoods, but their culture imprints these areas with their identities.
4. Enclaves are vibrant neighbourhoods with high home ownership rates and thriving commercial and cultural life. There are some rental apartment clusters of low-income immigrants.

Daniel Hiebert et al. did a comprehensive analysis of immigrants' enclaves in Toronto, Montreal, and Vancouver and they affirm these findings. They conclude that visible-minority groups reside "typically in multi rather than single ethnic areas and they are mixed in their socio-economic composition."[19]

Ethnic Enclaves in New York

New York is also a paradigmatic city of ethnic enclaves. This city of five boroughs had a population of 8.2 million in 2011, increasing by about 166,855 in the previous ten years. It has always been a city of ethnically/racially distinct neighbourhoods. Even in the early twentieth century, there were Irish, Jewish, Italian, and Black areas.

Map 4.2 shows the neighbourhoods where a particular ethnic group is numerically the largest in a census track, thereby making it an area of ethnic concentration. It is based on the American Community Survey data mapped by Ford Fessenden and Sam Roberts in a report in the *New York Times*.[20] Map 4.2 has reconfigured the mapped data according to its own categories. It shows the contemporary neighbourhoods where particular ethnic groups dominate. Dominicans, the largest foreign-born group in the city, have a large enclave in Washington Heights in Manhattan, spilling into the South Bronx, Puerto Ricans–Caribbeans dominate in East Harlem, and Hispanics in the Bronx, Corona (Queens), and Jackson Heights. Jewish concentrations are in Crown Heights and Midwood, Brooklyn, whereas Chinese and Koreans have enclaves in Forest Hill and Sunset Park, apart from Manhattan's historic Chinatown. There are many small areas of ethnic concentration, such as Russians in Brighton Beach, Brooklyn, Greeks in Astoria, Queens, and Pakistanis in Staten Island. New York's Department of City Planning observes that "immigrants' concentration in many neighbourhoods

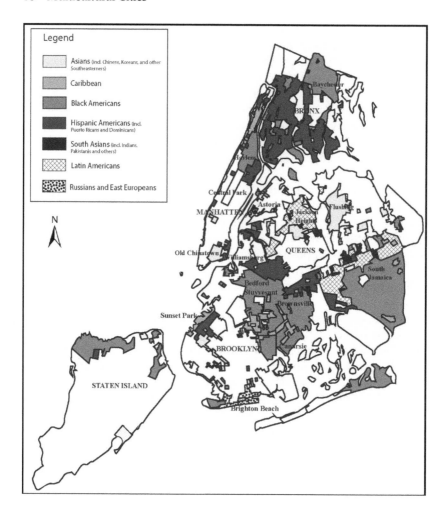

Map 4.2 New York's neighbourhoods by ethnic concentrations, 2010

Photo 4.2 Little Italy, New York (courtesy Milagros Dorregaray)

have resulted in ethnic enclaves, where an immigrant group leaves its social, economic and cultural imprint on a neighbourhood."[21]

New York residential geography changes relatively more quickly owing to the city's periodic property-market bubbles and mobility of residents. In the 2000s, Manhattan has become largely racially White. Even Harlem, the historic bastion of the Black community, is being gentrified and becoming mixed. The ethnic dynamics of New York is fluid, leading to the reconstruction of racial boundaries. Nancy Foner points out that New York is a place of "hybrid and fluid exchanges across group boundaries" that represents a new kind of multiculturalism.[22]

Like the enclaves in Toronto, New York's enclaves have also dispersed among them people of other ethno-racial backgrounds. For example, Flushing (Queens) is a new Chinatown, but it has Latin Americans, South Asians, Koreans, and native-born Whites as minorities.[23]

New York's ethnic enclaves bear many structural similarities to those of Toronto's (namely, suburban locus, domination of a group but not its exclusivity); but they differ in form, and meaning. The fluidity of the city's housing market affects their form, and race identity plays a greater role in the structuring of enclaves.

Ethnic Los Angeles

Los Angeles city has always been a gateway for immigrants, but now its suburbs in Los Angeles County are home to some of the biggest ethnic enclaves. Historically, the Los Angeles area has been a patchwork of low-density neighbourhoods and local municipalities knitted together by highways. The county has eighty-seven municipalities. These places have offered opportunities for home ownership at modest prices and thus have developed strong local identities. Immigrants moving into these areas imprint their identities on the landscape.

Map 4.3 has been extracted from the atlas of Southern California prepared by James P. Allen and Eugene Turner.[24] The Latino barrio in East Los Angeles, extending up to the San Gabriel Valley, is now the largest concentration of non-Whites in Los Angeles.[25] In the map, it forms a large ink spot in the middle. Mexican and Salvadorian immigrants "have superseded working class African American on the east side of South-central Los Angeles."[26] Latino barrios are strung along highways radiating out into the valleys. Even Watts, the historic Black American redoubt, has been Latinized, and middle-class Blacks are being transplanted to the west side.[27]

The emergence of Chinese ethnoburbs (suburban ethnic clusters of people and businesses) in Los Angeles County is a dramatic story of enclave formation. Monterey Park in San Gabriel Valley is an incorporated municipality, which has been remade since the 1980s with the arrival of investors from Taiwan and the Pacific Rim. Non-Hispanic Whites were reduced from 85% of the population in 1960 to 5% in 2010.[28] Despite Monterey Park being a Chinese ethnoburb, it had 27% Latinos and 19% other Asians.[29] Again, this is an affirmation of the non-exclusivity of ethnic enclaves. The Asianization of other towns such as Alhambra and Rowland Heights, forming an arch of Chinese, Korean, and Filipino business enclaves, has made the San Gabriel Valley a region of enclaves.

All in all, in both the city of Los Angeles and the surrounding county, social geography has been drastically transformed with the emergence of ethnic enclaves. The dominant trait of the area is its Latinization. But Latin areas are interspersed with Asian ethnoburbs and citadels of non-Hispanic Whites. Blacks seem to be squeezed territorially.

Los Angeles's enclaves have many similarities with those of Toronto and New York. They differ in terms of the sectorization of ethno-racial identities and the convergence of proximate ethnicities in enclaves.

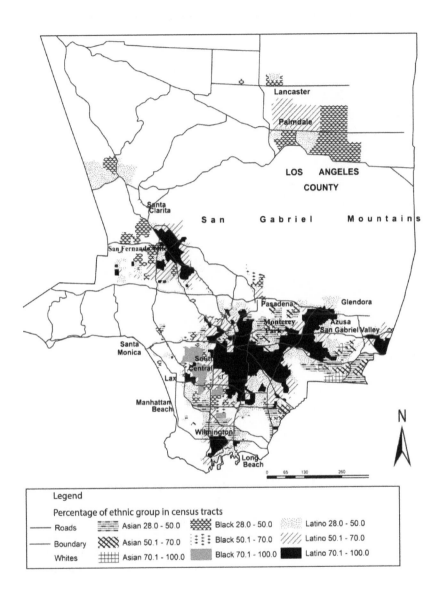

Map 4.3 Los Angeles County's ethno-racial concentrations, 2000 (from James P. Allen and Eugene Turner, *Changing Faces, Changing Places: Mapping Southern Californians* [2000])

Photo 4.3 Latino neighbourhood of East Los Angeles

Ethnic enclaves are emerging in most cities of North America. From Houston, Vancouver, and Miami to Dallas–Fort Worth, Phoenix, Sacramento, and Calgary, there are Chinatowns, Little Saigons, Indian bazaars, and Mexican barrios.[30] That they thrive is a mark of the multiculturalism of these cities.

Lest it be assumed that enclaves develop smoothly, a note of caution is in order. There are jurisdictions that resist the formation of ethnic enclaves. Also, turf battles develop among ethnic groups. In New York, Hasidic Jews and Afro-Caribbeans have clashed over the use of space and housing in the Crown Heights area of Brooklyn and Italians have been resistant to the Chinese moving into Little Italy.[31] Multicultural Toronto has had long drawn-out citizens' battles over the development of ethnic malls and places of worship. Monterey Park in Los Angeles County passed an English-only sign by-law in 1980 to keep out Asians. Yet such incidents do not seem to impede the demographic forces that drive the development of ethnic enclaves.

Global or Integrated Neighbourhoods

Ethnic enclaves represent cultural diversity across neighbourhoods, but they are not a majority of the neighbourhoods in the three cities. More common are neighbourhoods where many ethnic groups live together. They are multicultural at the fine grain. These multi-ethnic and multi-racial communities have been called global or integrated neighbour-hoods.[32] They can be formed in many ways.

People of diverse backgrounds choose a neighbourhood on account of its location or housing vacancies. This sets in motion a process of the infiltration of a new ethnic group into an established neighbourhood, leading to population turnovers. Also new housing built on vacant land attracts new groups.

In Toronto, global neighbourhoods are found in the gentrifying parts of the central city, such as Cabbagetown and Leslieville, in the expanding downtown, and in suburban Mississauga, Markham, or Scarborough. Public housing complexes usually have very mixed populations. For example, Toronto's largest public housing complex in the Jane-Finch area has 138 home languages. In Montreal, Côte des Neiges is the quintessential global neighbourhood, being an area of many nationalities.

In New York, the 2010 census shows that 1103 integrated neighbourhoods, where Whites and one or two minorities constituted 10%–20% of residents each, formed the majority of the city's neighbourhoods.[33] Ethnic enclaves realign the urban structure, but almost every part of a multicultural city has ethno-racial diversity.

Ethnic Malls, Plazas, and Economic Niches

Other manifestations of multiculturalism in a city's spatial organization are ethnic malls, plazas, and economies. These are commercial clusters producing and selling goods and services marked by the culture of a particular ethnic group.

An ethnic mall is usually an enclosed commercial centre with vendors specializing in the products and services of a particular ethnic group, such as groceries, bakeries, restaurants, video and book stores, clothing boutiques, travel agencies, and gift shops. Often there are immigration consultants, doctors and dentists, and real estate agents operating in ethnic languages. Such malls often do not have department stores or grocery chains as anchors, though a banquet hall or large clothing store may serve the same purpose. The striking quality of an ethnic mall is its distinct ethnic identity represented in goods, signage, the language spoken in stores, and decorations.

An ethnic plaza is usually an open-front cluster of ethnic shops mixed with a Seven-Eleven or Mac's store. It is usually small and often in the form of a strip mall along a major road.

An ethnic economy is essentially a network of commercial/manufacturing activities in which an ethnic group comes to specialize. It may or may not be spatially concentrated. It may not even deal in ethnic products. It could be a slice of the mainstream market in which an ethnic group comes to dominate such as Korean flower shops in New York or Sikh airport taxi/limousine services in Toronto. The Cuban business network in Miami is a famous example of an ethnic economy.[34] I will have more in-depth analysis of ethnic economies in the next chapter. Here I am concentrating on the spatial manifestations of ethno-racial economic activities.

A city's commercial sector is organized in a hierarchical order. Stores that sell products for daily use are found on local streets and form the lower levels of a hierarchy. Those selling goods that are relatively expensive and of periodic use, higher-order goods, cluster in central locations of districts, forming a second or third tier of a commercial structure. Luxury goods or special products are to be found downtown or main avenues in centrally accessible locations.[35]

In the suburbs, this hierarchy is reflected in the planned neighbourhood in community and regional malls, with the recent emergence of power centres as supra-regional commercial centres. Ethnic malls and businesses are injected into this hierarchy, serving both neighbourhood and community levels, on the one hand, and as special commercial areas of tourist and citywide interest, on the other. They reorder the hierarchy and scramble the boundaries among various levels of commercial centres.

Apart from these more recent forms of ethnic commercial centres, there are the historic ethnic business areas that evolve incrementally within old ethnic neighbourhoods. Chinatown is a universal example of a special commercial area in cities of North America and Europe. Of course, there are also Jewish, Italian, Greek, or Middle Eastern commercial areas in many cities.

Ethnic malls serve different functions for different groups. For ethnics, they are commercial centres where they buy groceries, clothes, videos, and CDs from the home country or visit a doctor, albeit to meet everyday needs. They also visit malls to eat, be entertained, and find pleasure in a place affirming their heritage. Others, including tourists, come to ethnic malls and commercial areas for fun and variety. An ethnic commercial area offers a quick excursion to "foreign lands" without going abroad. It also caters to the global tastes of contemporary city dwellers for the foods, music, and clothes of distant lands.

Ethnic malls have spread all across North America since the 1980s, riding on the wave of recent immigration. They are mostly Chinese, Korean, Japanese, and, recently, Vietnamese in character, though Indian/Pakistani and Mexican/Colombian plazas have also begun to appear. Los Angeles incubated this trend. It was the first city to have Korea and Japan Towns as malls in distinction to the historic Chinatowns or Korean commercial street. In a short span of time, Asian malls have followed wherever large concentrations of South East Asians settled.

Asian malls are found not only in the immigrant-rich cities of California and New York, but also in places such as Atlanta, Houston, Washington, and Denver. Asian Square Mall is part of a six-mile stretch of ethnic businesses along Buford Highway in Atlanta. New York City's Chinese Mall in Flushing is paralleled by Colombian strips on Roosevelt Avenue, the Indian commercial cluster in Jackson Heights, and the Russian shopping arcade in Brighton Beach, Brooklyn.

Perhaps the most striking case is that of Toronto, where, in an area enclosed by Leslie Street in the west and Markham Road in the east, and by Sheppard Avenue in the south and Highway 7 in the north, there are fifty-nine Chinese malls and plazas. In the Toronto CMA, there are sixty-six Chinese commercial clusters.[36] One of the largest North American Asian malls, the Pacific Mall/Market Village in Markham, has three hundred Hong Kong–style stores and restaurants. Another bigger mall is being developed next door. These malls have all appeared since 1980, and complement three other Chinatowns in the Toronto area. Vancouver's Crystal Mall in Burnaby and the Aberdeen Centre in Richmond are other examples of famous Asian malls in Canada. Similarly, Indian/Pakistani plazas and malls have appeared in Toronto and Vancouver.

As the commercial hub of ethnic life in cities, ethnic malls and shopping areas are not only growing but also evolving in parallel with the changing demands of settled immigrants and the second/third generations of ethnics. They are diversifying the mix of stores and merchandise, as demand changes with second and third generations of immigrants. A relatively new trend in Asian malls is the offering of regional goods and services rather than just those of one nationality. Instead of being made up of primarily Chinese stores, new Asian malls mix them up with Korean, Japanese, and Vietnamese boutiques and restaurants. The next stage is probably the development of multi-ethnic malls where stores of a wide range of nationalities could be brought under one roof. This trend is already reflected in food courts and on ethnic commercial streets where West Indian groceries, Colombian bakeries, and Indian sari shops, for example, are located side by side.

All in all, a robust ethnic commercial sector has emerged in multicultural cities. It has reordered the commercial hierarchy and transformed cities into polynodal places. Instead of a single dominant centre, multiple focal points have emerged, affecting the role of the downtown. Ethnic malls and commercial areas bring variety and colour to the homogenized landscape of suburbia.

Before we conclude the discussion of ethnic malls and commercial areas, it should be pointed out that these developments have not come about smoothly. What has been described above is the current state that has resulted from almost thirty years of gradual erosion of resistance to the attempts of ethnic minorities, particularly non-Whites, to express their identities. In almost every city, the first attempts to build ethnic malls or develop places of worship were met with strong opposition from local communities. From stereotyping to pressuring city government to reject proposals to build an ethnic mall or place of worship on the basis of their supposed adverse impact on a neighbourhood, obstacles were put up to resist such developments. As precedents were established in a city, subsequent developments had a smoother passage. Every city has, by and large, gone through a social learning process. It is to the credit of democratic citizenship and the regime of civil rights that in many North American cities of large immigrant populations, ethnic establishments and institutions have come to be accepted, signalling the arrival of the multicultural city.

Synagogues, Mosques, Mandirs, and Gurudawaras

Non-Christian places of worship are another expression of multiculturalism in an urban landscape. To the extent that non-Christians succeed in building such facilities, they are successful in integrating into a city. Non-Christian places of worship represent new elements of urban structure and stand as an emblem of cultural and religious diversity. They have significant impact on the spatial organization of a city.

Historically, churches were functionally and aesthetically an anchor of urban life. Church spires dominated a city's skyline, and its centre

Photo 4.4 Bangladeshi Mandir, New York (courtesy Milagros Dorregaray)

revolved around the cathedral on the main street. Churches themselves are not of one design or denomination. The onion domes of an Eastern Orthodox church could not but offer a contrast with the soaring gabled roof of a Presbyterian church. The spread of cities and dispersal of congregations to the suburbs have eroded the role of neighbourhood churches, as has the secularization of social life. In the centres of cities, the building of high-rises and the emergence of gleaming stores, clubs, and bars have eclipsed cathedrals and churches as architectural landmarks.

Jewish synagogues initially injected a note of religious diversity in urban places of worship. They introduced the first wave of non-Christian religious architecture. Though at first limited to the central city's Jewish neighbourhoods, they can now be found in suburbs where younger generations of Jews have moved. The suburbanization of synagogues has cast their architecture in the international style. Large, high-ceilinged and barn-like structures, surrounded by large parking lots, blend into the suburban landscape. Synagogues essentially fall within the architectural idioms of Western civilization.

The arrival of new citizens from Asia and Africa has introduced new faiths – Islam, Hinduism, Buddhism, Zoroastrianism, and Sikhism – and newer forms of Christianity, that is, Chinese, Korean, and African. With these faiths have come not only new types of places of worship introducing a wide range of architectural and institutional idioms, but also new forms of worship. The diversity of forms and functions is striking.

The development of new places of worship, namely, mosques, mandirs, gurudawaras, and Buddhist temples, has followed the usual path of community opposition, resistance, compromise, precedents, and acceptance. When these institutions are first proposed in a city, they are often noisily opposed on the basis of their "alien" activities, unfamiliar architecture, and presumed incompatibility with the surroundings. Nimbyism is aroused, with a subtext of ethnic and racial prejudice. As the first such institutions are built and the ranks of potential voters swell with immigration, local authorities begin to modify their development regulations to accommodate non-Christian places of worship. This scenario has been repeated, more or less, in all cities.

After the 9/11 attack on the World Trade Center, proposals for Islamic centres and mosques have aroused stereotypical apprehensions about terrorism and public safety. Switzerland's People's Party, a mainstream but right-wing national party, banned minarets in mosques in 2008 on the ground that they are intrusive in a secular country. Similar controversies have erupted in Italy and Belgium. In the United States and Canada, such controversies are largely local and are expressed through the citizen-participation processes of city planning. Yet it is a credit to the responsiveness of the process that almost every big city in North America has now hundreds of mosques, mandirs, temples, and gurudawaras, not to mention Korean, Chinese, and Ethiopian churches.

The Toronto metropolitan area has about one hundred mosques, seventy temples and nineteen gurudawaras. Though many of these institutions are housed in adapted homes, stores, and warehouses, close to a third or more are newly built, architect-designed structures. The Swaminarayan mandir cum Hindu museum in Vaughan, made of sculptured Turkish limestone and Indian marble, is now listed as a place of interest for visitors to Toronto. Looming over the intersection of two major highways (401 and 400) is the Taric mosque, whose resplendent Arabic calligraphy catches the eye of commuters zipping across the north of Toronto. The golden dome of Ontario Khalsa Darbar is a striking presence on Dixie Road in Mississauga. These examples illustrate the extent to which Islamic, Hindu, and Sikh places of worship have been woven into the spatial fabric of the Toronto area. This has come about in the way described above, graduating from resistance to integration.[37]

In New York City, there are seventy-five mosques, thirty in Brooklyn alone, thirty mandirs, and numerous gurudawaras and Zen Buddhist temples. Los Angeles County had fifty-nine mosques in 2010, increasing by eleven since 2000.[38] All in all, there are approximately 1200–1500 mosques in the United States, and new ones continue to be developed. These snippets of evidence testify to the incorporation of non-Christian places of worship in the urban landscapes of North America.

Most of these places of worship are to be found in the suburban parts of metropolitan areas. This is where immigrants are settling. Yet they seldom re-enact the role of neighbourhood churches of the bygone era. They draw their congregants from far and wide and not primarily from the surrounding neighbourhood, though sometimes the establishment of a mosque, mandir, or Eastern Orthodox church brings people of that faith to settle in the area.

Of course, there are such places of worship in the central areas of cities, but these are largely housed in storefronts, abandoned churches, converted homes, or warehouses. Central cities are built up, their land prices are high, and lots are usually too small to meet zoning requirements. All these factors combine to promote the development of these

Photo 4.5 Mosque, Toronto

new forms of places of worship in suburban areas and they are continually being driven to the periphery in search of cheap land and large lots
to build sizeable establishments.[39]

Another characteristic of non-Christian places of worship is that,
within the same faith, they are differentiated by ethnic and denomi-
national differences. Ethnic imprints on places of worship arise from
sponsoring communities. Somalis manage this mosque and that one
is of Punjabis. One mandir may be devoted to the worship of Hanu-
man, while another may be dedicated to Vishnu; Gujratis may manage
one and another may be serving Indo-Caribbeans. The prayers may
be similar but the sermons, language, and social activities are often
different.

There is a functional evolution in new places of worship. They are
also becoming community centres where a variety of programs are
offered, including secular activities such as presentations on immigration laws or heart health and employment counselling.

All in all, non-Christian places of worship and non-Western churches are a vivid symbol of multiculturalism and the religious rights of minorities. They add another layer of activity nodes to the spatial organization of a city. They also create different daily, weekly, and yearly rhythms of activities. These differences, arising from a variety of identities and beliefs, inject new circuits of movement, symbols, and architectural idioms in the urban landscape.

Public Space and Diversity

A city is knitted together by its streets, roads, parks, buses and subways, schools, stadiums, monuments, halls, museums, art galleries, and other public places. These are the venues where people come together to see and meet one another. This is where the civic spirit and a shared destiny are forged. Richard Sennett lauds the virtue of exposure as the mechanism by which an individual experiences a city and finds his or her orientation.[40] These places are subsumed under the term public space, which stands in contrast to the private space of home, store, workplace, and church. Public spaces promote civic culture by establishing norms, values, and ethics of behaviour in such places.

Public space is encoded with the culture(s) of a city. Historically, it is the culture of the politically and economically dominant segments of the local population. Multiculturalism opens the way for minorities' cultures to be expressed in the public space, and through mutual exposures renew civic spirit; that is how a city serves as a common ground.

In multicultural cities, parades, fairs, and protests are one expression of cultural diversity and, of course, an affirmation of democracy as well as minorities' citizenship rights. An indication of the scope of cultural diversity expressed in the public space in New York City, can be found in a listing of one month's parades: they include those for Memorial Day, the Golden Marijuana March, Cuban Day, Turkish American celebrations, Haitian Flag Day, Norwegian American 17th of May, the Salute to Israel, and the Greater New York Good Neighbour event. The City of Toronto had 310 street closures for parades and community events in 2002. Its Gay Pride Parade, Caribbean Festival (formerly Caribana) West Indian Parade, and Santa Claus Parade are events that annually draw thousands of participants and a million as spectators from distant places. New York's St Patrick's Day Parade is now rivalled by the West Indian parade in numbers.

International events also reverberate in multicultural cities. Be it the Palestinian-Israeli conflict, famine in Africa, an earthquake in Pakistan, or the US presidential election, these issues are aired in cities in the form of public meetings, marches, and protests.

Apart from fairs and parades, the sights, sounds, and smells of the public space are transforming multicultural cities. Stores with Chinese, Arabic, or Spanish signs, for example, line streets not only in ethnic neighbourhoods but also in city centres. Mannequins draped in saris in a store window, glazed Peking ducks dangling from hooks beside a restaurant's entrance, souvlaki being cooked on food carts in a plaza, an Ecuadorian band playing on a sidewalk, or herbs and spices piled high in bags lining a storefront are some of the sights and sounds that not only bring colour and variety to public places, but also reflect the extent to which cultural diversity has seeped into the public space. Perhaps nowhere has multiculturalism taken hold as strongly as in the food courts of shopping centres all over North America.

Similarly, bilingual street signs, the strains of plaintive music, and arches and bunting marking the territorial identities of ethnic enclaves inject a new feel and look into public places. At the level of individual properties, small expressions of residents' tastes in decorations, lawn furniture, and the colours of façades help to create a great variety of streetscapes. A statue of Venus beckoning from a lawn, a mosaic of Rastaferians staring down from the side of a house, a front lawn turned into an herb patch or an Arabic blessing painted above a front door, for example, are small statements of multiculturalism.

The Social Contours of the Multicultural City – Spatial Segregation and Assimilation

The spatial expressions of ethnicity do not add up to making a new city. They essentially graft new activities, practices, and symbols onto the existing structure of a city. These elements create new nodes of activities and interests, generating new focal points and realigning urban structures.

The layout of a city, its organization into commercial, industrial, and residential districts, and its system of streets and roads are etched on the ground. They have evolved over a long period under the influence of historical, technological, social, political, and economic forces. Cultural diversity works within these structures. It modifies but does not eradicate them.

The racial and cultural pluralism of a city has been viewed through the lens of ethnic and racial segregation that has long characterized the neighbourhoods of minorities, particularly of Blacks in the United States. Residential segregation and the spatial assimilation of minorities are two themes that have preoccupied the social geography literature. Residential and economic segregation resulting from discriminatory laws and practices is undoubtedly an intolerable outcome. But if they are the product of social choice and community development, is that still socially detrimental and morally abhorrent? This question underlies the current discussions about ethnic enclaves, economies, and institutions in North American cities.

Douglas Massey and Nancy Denton express the spatial assimilation of minorities as a function of their socio-economic advancement and improved housing situation, which "implies assimilation with [the] majority group."[41] These ideas predate the contemporary ideology of pluralism and the new demography of cities. Assimilation does not seem to be the ladder for ethnics to ascend to higher social strata. An illustrative example is that of the Jewish, Italian, and Portuguese enclaves in the Toronto area. These are enclaves of long-established and prosperous ethnics. Many ethnic enclaves thrive on the prosperity of immigrants, who as investors and professionals enter directly into the middle and upper echelons of society. They help build ethnic enclaves and institutions as the geographic expressions of diversity and evidence of a city's cosmopolitanism. This process has brought about suburban enclaves, architecturally striking places of worship, and community institutions.

Residential segregation has long been associated with racial discrimination. It is particularly true of the US cities where poor Blacks have long lived in slums, neglected by cities and isolated from society. Yet residential and commercial concentrations of ethnics and immigrants seldom rise to the level of segregation typical of ghettos. First, their enclaves are seldom exclusively of one class and ethnicity, though a particular group may dominate. Second, these are places of well-functioning community organizations and institutions, not infrequently of solid socio-economic standing. Even in the poor immigrants' neighbourhoods, educational and other social indicators are comparable to that of the mainstream. These are places of hope and not despair.

Canadian studies are almost unanimous in rejecting the notion of ethnic enclaves being ghettos and neighbourhoods in decline. The evidence in Canada is much more clear that ethnic enclaves do not isolate immigrants and are not a transitory phenomenon.[42] R. Walks and Larry

Bourne conclude that the "association of low income and high level of minority concentration holds only for some urban areas and some minority groups ... [There is] little evidence of ghetto formation."[43] Daniel Hiebert concludes from a longitudinal analysis of the visible-minority enclaves in Montreal, Toronto, and Vancouver (1996–2006) that they are "characterized by profound ethno-cultural diversity and [the] propensity to be poor in these enclaves is not higher than that of those who live outside the enclaves, by and large."[44]

There is almost a transnational consensus among geographers and sociologists that contemporary enclaves are not ghettos in the sense of being haunts of poverty and discrimination.[45] John Logan, Wenquan Zhang, and Richard Alba make a distinction between an immigrant enclave associated with labour migrants and an ethnic community of entrepreneurs and professional immigrants. They found that these two types of ethnic neighbourhoods in New York and Los Angeles include groups that continue to live there even when persons of similar socio-economic backgrounds have dispersed. Los Angeles had a more decentralized form of ethnic distribution, whereas New York was more centralized. Yet in both cities, parallel evidence for ethnics' self-selected separation as well as assimilation was to be found.[46] There were differences among ethnic groups as well as between New York and Chicago, on the one hand, and Los Angeles and Miami, on the other.[47] These studies point out that the residential and economic clustering of immigrants and their progenies are a multidimensional phenomenon and cannot be reduced to the one-dimensional narrative of segregation-assimilation. What we see on the ground in multicultural cities are the manifestations of thriving ethnic places both geographically and socially. This is a new form of city. It bears repeating that a majority of ethno-racial groups live in integrated neighbourhoods, where many groups live side by side.

Post-modern Currents in Cities

A city of ethnic spaces has been called, variously, an EthniCity, post-modern city, intercultural city, or fragmented city. Strictly speaking, each of these terms claims to capture the reality of the emerging multi-cultural and post-industrial city, but apart from an academic proclivity to coin new terms, each approaches the phenomenon from a related but different angle. "EthniCity" as a term focuses on the cultural diversity of the residents and wide variety of people making up the population of a city, capturing its multi-ethnic character.[48] "Intercultural city" is more of an ideal type than a description of an existing place. It envisions a community where minorities and the majority publics are organizationally and socially linked together in a public sphere that reflects their mutual interests, not unlike the idea of common ground.[49]

"Post-modern city" as a model explains the role of ethnic spaces in restructuring a city by linking them with the emergence of information technology and global production and trade networks. The Los Angeles School of urban geography advances the idea of a post-modern city. Its arguments are complex but the basic idea is that production processes have been transformed by electronic technologies, allowing firms to specialize in a narrow range of activities, by contracting out and externalizing many of their functions, and thereby seeking horizontal integration by clustering near those in the same line of production.[50] From this dual process of fragmentation and specialization of functions emerge labour market niches in which different ethnic groups tend to dominate, giving rise to ethnic networks and enclaves.[51] The post-modern city lacks a dominant centre. It is fragmented spatially and is tied together by communication networks. Ethnic spaces of a distinct character are elements of such a city.

To conclude these arguments, it can be said that ethnic spaces are not an isolated phenomenon driven only by processes of social integration or segregation. Rather, they represent transformative trends in the structuring of cities arising from changing demography, technologies, global trade, immigration, and the revolution of "rights" that sustain ethnic cultures and identities. They represent new socio-spatial forms grafted onto the structure of a city.

The City as a Common Ground

Ethnic enclaves, economies, and institutions are not independent parts of a city. There is a city underneath and around them that sustains them and defines them. Its history, economic organization, political institutions, infrastructure, and civic culture underpin ethno-racial diversity. The ethnic spaces of a city are embedded in its roads, utilities, public transport, traffic regulations, land use policies, and other collective goods. Can one imagine an ethnic enclave that is not connected with a city's downtown, or its truck and rail terminals, or covered by public health or police services, for example? This is how a city shapes the form and structure of ethnic spaces. The functioning of these spaces also requires common norms, values, and behaviours – shared civic culture.

Urban living entails some functional imperatives that permeate all elements of a city. Among these imperatives are (1) collective goods; (2) interdependencies of land uses and externalities-based property values; (3) local government and civic culture; and (4) urban social organization based on secondary relations and impersonally linked roles. They affect everybody and influence the organization of all communities. These imperatives cultivate a civic culture of common public behaviours, values, citizenship, and ethics. This is how a city becomes a part of the common ground which defines, accommodates, and harmonizes diversity. A multicultural city has a pluralistic civic culture. It is a process more than a condition.

Ethnic communities may have strong bonds and dense social networks, but their community life is also bound together by telephones, ethnic newspapers, websites, Facebook, associations, and institutions, because face-to-face relations have limited purchase in communities of thousands of members. Ethnic solidarity operates as one among many interests.

These imperatives of urban living underlie the ethno-racial geography of multicultural cities. They affect the social sustainability of ethnic and racial communities as much as the city as a whole. Mario Polese and Richard Stren define social sustainability as "policies and institutions that have the overall effect of integrating diverse groups and cultural practices in a just and equitable fashion."[52] These are the prerequisites of a multicultural city. They make it possible for diverse groups to thrive, while linking their well-being to the welfare of all. Polese and Stren conclude from case studies of cities ranging from Toronto, Miami, and Nairobi to Cape Town that policies promoting equitable development of urban infrastructure and services are instruments of social sustainability and integration.[53] Through these institutions and practices, the city becomes a significant element of common ground that sustains diversity.

Conclusion

Multiculturalism wedges new activities, communities, and aesthetics into the spatial structure of a city. Within the historic structure of a city, namely, its downtown, commercial districts, rich and poor neighbourhoods, office towers, university campuses, and open spaces, multiculturalism injects some new functions but mostly gives new forms to existing functions. A neighbourhood remains a residential area, though now it could largely be the home of a particular ethnic group. An ethnic mall, though of the size of a community shopping centre, offers specialized goods and services that are of a regional order. All in all, multiculturalism realigns the spatial structure of a city, while retaining its historic contours. It also frames new developments in cities.

This chapter shows that the social geography of multicultural cities, be it Toronto in Canada or New York and Los Angeles in the United States, takes similar spatial forms, namely, ethnic enclaves, places of worship, ethnic economies, and community institutions. Obviously, the same ethnic group in different cities has similar needs and meets them in parallel forms. Ethnic enclaves have now spread out to the suburbs in the three cities. They are large, relatively prosperous, dominated by a particular ethnic group, and have similar aesthetics and activities, but not without the internal diversity of class and nationality and the presence of proximate ethnic groups. The same is true of other spatial and physical facilities and institutions. Yet a majority of neighbourhoods are integrated, including two or more groups mixed together.

The city injects some uniqueness in functional role, internal organization, and the locations of various places and institutions. The differences arise from the historic urban structure, housing and commercial markets, and civic cultures. New York, Los Angeles, and Toronto differ in terms of the meaning and the economic role of their multicultural places. Race has a stronger role in defining the social geography in US cities than in Canadian cities, where ethnicity and national origins carry more weight. The ethno-racial patterns of US cities are fluid and hybrid, whereas Canadian multiculturalism tends to embed ethnic identities more enduringly, as in the older ethnic enclaves of Italians, Jews, and other Southern Europeans that have lasted from three to four generations despite their evolution.

Multicultural cities are the product of the accommodations and harmonizations of differences. They are continually reconstructing civic culture through evolving public norms and values. They are exciting places but not without the tensions of accommodating diverse cultures. There will be more about these processes in later chapters.

Chapter Five

Ethnicity and the Urban Economy

Ethno-racial Alignments of the Urban Economy

Racial and ethnic differences are reflected in urban economies, in their institutions of production and distribution as well as in the incomes and consumption of different groups. The divide between Whites and non-Whites in employment and income is a long-standing racial contour of the American and Canadian economies. The arrival of a large number of immigrants of diverse ethnicities and the growth of their native-born second/third generations in the past half-century have injected new differentiations on the basis of ethnicity, culture, nationality, and religion in economic organizations.

There are differences in consumption needs and tastes by race, ethnicity, and (sub)culture, which translate into variations of products and the profiles of demand. Correspondingly, employment opportunities, human and financial resources, social and cultural capital, and enterprise also vary by race, ethnicity, and immigrant status. Overarching these factors is the institutional structure of public regulations, policies, rights, and laws, which organize economic transactions. These factors lead different groups to concentrate on particular occupations and activities, and spawn ethnic economies, networks, and niches organized around their cultural practices and social organization. These are the ethno-racial contours of urban economies.

This chapter aims at probing the patterns of economic differentiation by race, ethnicity, and immigrant status and observing the ethno-racial alignments of the urban economies of Toronto, New York, and Los Angeles. Its focus is on those aspects of the urban economies that are affected by the cultural and racial diversity of economic actors. It does not aim at giving a comprehensive account of the three cities' economies. However, the specific questions probed in this chapter are: How are ethnic economies organized; how do they evolve and what role do they play in area-wide economic organizations? What factors promote ethnic and immigrant entrepreneurship? How do those factors play out in the three cities? Finally, how do different groups fare in realizing growth and equity in the distribution of opportunities and incomes?

Ethnic and racial differences in an economy are manifested in two ways, as vertical divisions within economic sectors and as horizontal clustering in selected activities and occupations. The concentration of minorities and immigrants in the lower strata of economic sectors is a common vertical contour. Historically, discriminatory laws and practices confined ethno-racial minorities to the lower-tier occupations, such as labourers, domestics, farm and industrial workers, or clerks. Women also confronted similar barriers. The enforcement of civil and human rights and employment-equity laws has reduced discriminatory practices, but minorities and immigrants continue to have a disproportionately large share of the low-level jobs. Apart from the historical legacies, this concentration in the lower tiers of economies is also the result of immigrants' youthful profile in the labour force and their inexperience in local labour markets. Of course, now immigrants and minorities have greater mobility, and the entry of the appropriately qualified to professional and executive positions is relatively open. The vertical clustering turns into social-class differences within ethnic and racial groups.

The horizontal clustering in particular occupations by ethnicity is another form of the ethno-racial contour of urban economies, for example, the concentration of Chinese, Indians, and Russians in information technology, engineering, and finance. Ethnic concentration in businesses producing ethnic goods and services, for example, Kosher foods or Indian clothing, is another form of clustering.

Overall, ethnic and racial clustering in some occupations, activities, and lines of business falls into a range of organizational patterns, from a thin ethnic concentration in a sector to economic networks and (sub) economies structured around shared identities and cultural values. In such economic clusters or ethnic networks, special talents, community solidarity, and culture are the organizing factors. These economic clusters are not outside the mainstream economy. They are embedded in national and urban economies, and are the most striking manifestations of ethnic and racial contours in urban economies. I will be mostly focusing on them in this chapter.

Types of Ethnic Economies

Culture matters in economics, as does place, even though much of the economic literature ignores it. An economic organization itself is a cultural artefact based on a network of institutions organized around norms of exchange, demand satisfaction, property rights, business ethics, enterprise, and transactions. Within societies, these norms vary a bit by class and ethnicity. For instance, prices in stores are viewed differently in various cultures. In Anglo stores, they are usually non-negotiable, but in Korean, Indian, or Latino stores they may be subject to bargaining. From pricing norms to financing practices, labour relations, and supply networks are a range of factors that vary by ethnic groups. These differences lie behind the clustering of production and distribution activities by ethnicity, nationality, religion, or language.

A useful way to begin defining ethnic economies is to give some illustrative examples. Chinatown is a ubiquitous example of an ethnic economy in almost every major city of North America. Miami's Cuban business enclave is another widely known example of an ethnic economy. Koreans dominating the flower trade or the historic concentration of Jewish entrepreneurs and workers in the garment industry in New York are other types of ethnic economy, which though geographically dispersed operate as a coherent network based on the trust of shared identity.

Ivan Light and Steven Gold define ethnic economy as a network of the self-employed, employers, and employees of an ethnic group.[1] Edna Bonancich and John Modell define it as an employment pool created by ethnics/immigrants.[2] This is essentially a labour-market perspective of an ethnic economy, which is more than a job pool of co-ethnics.

An ethnic economy operates on the basis of cultural knowledge, expectations of ethnic solidarity, and obligations of mutual trust. There are practices within ethnic economies that may not be common in the mainstream economy, such as character loans, personal dealings in preference to the impersonal transactions, or a sense of social solidarity. Business relations are buttressed by social ties forged through connections made at political, religious, and community events.

Economic dealings are embedded in mutual trust, common language, communal bonds, financial collaboration, and/or geographic concentration. This does not mean that dealings are not guided by self-interest, but that self-interest is pursued through shared strategies. All in all, an ethnic economy is more than a collection of ethnic businesses and their workers. It is also an integrated network of interlinked entrepreneurs as well as a segmented labour market for workers of an ethnic group, particularly immigrants of such backgrounds. Though an ethnic economy is operationally defined as a "set of businesses owned and operated by an ethnic group," it is also a social organization of mutual obligations and shared language and identity.[3] It could take a slice of the mainstream economy or specialize in the production and distribution of ethnic goods and services.

An ethnic-enclave economy is a cluster of ethnic businesses, which often has a territorial base, such as Chinatown, Little Havana, or a Latin barrio. The locational proximities foster interdependencies and agglomeration economies.[4] The territorial clustering of activities turns an ethnic economy into an enclave economy.

Light and Gold identified two other variations of ethnic economies: (1) an ethnic ownership economy and (2) an ethnic-controlled economy. An ethnic ownership economy refers to a network of small and medium-size businesses owned by ethnic entrepreneurs or immigrants of similar backgrounds.[5] A large body of literature has emerged about studies of the self-employment and business ownership of Chinese, Latinos, Cubans, Koreans, Blacks, and so on in the United States and Canada as well as of South Asians, West Indians, and others in Britain.[6]

An ethnic-controlled economy is another species of the ethnic economy, and refers to the phenomenon of a particular ethnic group having a strong presence in an industry or sector as owners or employees. It refers to the economic power of an ethnic group in an industry.[7] Ethnic-controlled economies do not necessarily specialize in ethnic goods. They are frequently engaged in producing some mainstream goods and services. Examples include Koreans in Los Angeles controlling soft-drink dealerships, Japanese growers as price determiners of strawberries in California, Italian builders in New York, or Sikhs dominating taxi services at Pearson Airport in Toronto.

The concept of an ethnic-controlled economy is synonymous with that of an ethnic niche in employment. An ethnic niche is formed with the special skills, resources, or networking power of an ethnic group in order to dominate an economic activity. Chinese herbalists, Cuban cigar-makers, Anglo bankers, or Jewish doctors in New York form ethnic niches in those industries.

The early formulations pointed out how ethnic economies operate on credit circles, personal connections, chains of mutually supporting deals, and old-world values.[8] Ethnic economies have been differentiated by their ability to leverage social relations and cultural mores into resources for business and self-employment. All in all, they were conceived by theorists to be rooted in social institutions to a greater degree than a modern economy.

Ethnic economies and niches are a distinct but an integral part of urban economies. They are subject to the recession and expansion cycles of the mainstream economy and are regulated by national and local fiscal, monetary, tax, and consumer protection laws, labour rights, and commercial-industrial practices and regulations. These are all elements of the common ground necessary for the sustenance of diversity within a society. The internal structure of ethnic economies is complemented by their external dependence on urban economies. Given that ethnic economies are products of immigrant (ethnic) entrepreneurship, they have the characteristics of what Robert Kloosterman et al. describe as mixed embeddedness. They are "located at the intersection of changes of socio-cultural frameworks on the one side and transformation processes in urban economies on the other."[9]

The Evolution of Ethnic Economies

Ethnic economies have historically begun as constellations of family restaurants, groceries, and other businesses meeting the daily needs of immigrants. As the ethnic population increases, the demand for ethnic foods, music, clothes, services, and so on expands; it spawns wholesale, import, and manufacturing activities.[10]

The second stage in the evolution of ethnic economies comes with ethnic businesses branching out into the mainstream economy, producing goods and services for other communities and building on the advantage of ethnic resources and networks. The third stage consists of the emergence of ethnic financial institutions, banks in particular, and the arrival of immigrant investors, entrepreneurs, and professionals who carve ethnic niches in the mainstream economy, bypassing historic ethnic enclaves. They forge transnational linkages and turn ethnic economies into platforms for global trade and production. Monterey Park in Los Angeles County is an example of an ethnic economy based on foreign investments and global links.[11]

How ethnic economies evolve is illustrated by the example of the Cuban enclave economy in Miami. The ground for the ethnic economy was laid in the 1950s in a poor neighbourhood of Miami with small stores started by newly arriving Cubans. By the 1970s, second- and third-generation Cubans had implanted a nucleus of corporate and professional establishments in Coral Gables, operating in both the United States and Latin America. Almost in parallel, a complex of workshops and industries specializing in garments, electronics, and construction emerged in Hialeah drawing on the capital and labour of a growing Hispanic population of South Florida.[12] The ethnic solidarity that spawned Little Havana flowered in Coral Gables and Hialeah. The Cuban ethnic economy has increasingly integrated into the regional economy, while maintaining its ethnic roots.[13]

The transnational linkages of ethnic businesses are contributing to connect mainstream economies with opportunities abroad. Salvadorian and Colombian companies run transport businesses in Latin America from Miami, Los Angeles, and New York. Dominicans in New York invest in properties on the Island.[14] Taiwanese and Indian computer chip designers manufacture their products in China and India. Toronto has an actors' academy for aspirants to roles in Bollywood movies in Mumbai. Robust TV, radio, and music industries emerge to serve a multicultural public. Ethnic newspapers, magazines, radio, and television become a link between the local ethnic market and businesses in the home countries. Multi-branch international corporations operating globally are increasingly tapping ethnic business networks to produce and distribute products. International banks are branching into Islamic Sharia-compliant banking outlets.

The Anatomy of Ethnic Entrepreneurship

Ethnic economies are an expression of the entrepreneurship of immigrants and its cultural carry-over to their second and third generations. What is the basis of this entrepreneurship? This question has long engaged theorists. There is a tradition in social sciences to explain the entrepreneurial behaviour of an ethnic group in terms of its cultural-religious values and social institutions. This approach is called group characteristics.[15] Max Weber's thesis of "the Protestant ethic and the spirit of capitalism" has long ruled over the theories of entrepreneurship.[16] Similarly, the fact that Lebanese, Greeks, Syrians, and Gujrati-Ismailis have dominated local trade in Africa, the Gulf States, and even some parts of the Caribbean has given rise to notions of their cultural proclivity for business. They have been given the name "middleman minorities" in contemporary economic sociology. Cultural theory attributes the entrepreneurial propensities of an ethnic group to such values as risk tolerance, family solidarity, frugality, hard work, adventurousness, and faith in one's destiny as well as community obligations. It is presumed that some cultures cultivate these values and habits more than others.

Another theory explaining the entrepreneurial propensity of some ethnic groups is called opportunity structure or the structural approach.[17] It suggests that entrepreneurship is a matter of market conditions or opportunity structures. Immigrants finding themselves as a minority among strangers seize unmet needs in ethnic markets and create opportunities for ethnic niches in the mainstream economy.[18] Native-born ethnic minorities also resort to similar strategies. It is the market that provides openings for the exercise of entrepreneurial talents. For example, as the number of immigrants increase, opportunities open up for starting restaurants, groceries, travel agencies, video stores, and ethnic newspapers as well as TV shows. These holes in the market invite entrepreneurship.

Yet another proposition is that ethnics' and immigrants' entrepreneurship, reflected in their high rates of self-employment, is a response to the blocked opportunities or barriers they face in the job market. Historically, Blacks, Chinese, Japanese, Jews, and Italians, in the 1930s and 1940s, were barred from professional jobs and faced difficulties in finding employment in the United States and Canada. Immigrants even now find that their professional credentials are not recognized. For example, the degree of an engineering graduate from Egypt may not be accredited for professional work nor is her Middle Eastern experience considered relevant in Canada and the United States. Such conditions leave no choice for immigrants but to resort to small businesses for their livelihood. The story of a physician driving a taxi or an architect running a dollar store is a common tale in immigrants' narratives.

These three seemingly different explanations of ethnic entrepreneurship have been combined in what is called the interactive model, as conceptualized by Roger Waldinger et al.[19] They maintain that an ethnic group's entrepreneurial success is the result of ways in which opportunity structures combine with the group's resources and characteristics. A group's internal factors are combined with external opportunities to form ethnic strategies, consisting of plans and actions, to realize the

Table 5.1 Bases of entrepreneurship

Resources and characteristics	Opportunities and market conditions
Class and human resources	*Local and regional economy*
Age, gender	Market conditions
Socio-economic status	Ease or difficulty of entering market
Assets and income	State policies – taxes, licences, credit
Profession	Local laws, zoning, infrastructure
Education and skills	Citizenship and immigrant rights
Attitudes and work ethic	Demand for ethnic goods
Family structure	Labour market and job opportunities
Cultural capital – manners, speech	Concentration of ethnic population
Human relations	Free trade and globalization
Ethnic (cultural) resources	
Cultural values	
Community support	
Business traditions	
Social capital	
World view, symbols, and identity	
Ethnic solidarity	
Language(s)	
Religious and moral orientation	
Transnational connections	

potential of entrepreneurship. The factors that promote entrepreneurship are sketched in table 5.1.

The table shows two types of resources and characteristics having a bearing on the ability to initiate and manage a business or project. One type is class and human resources that relate to the talents and qualities of an individual, though some of these may be drawn from the social structure in which the individual is embedded. Included in these are age, gender, social class, financial and physical capital, education, manners, and other elements of human capital.[20]

The second type of resources are qualities rooted in a person's ethnic and cultural background.[21] Such knowledge and practices can confer advantages in relevant lines of business. Ethnic resources also include social networks, community solidarity, business traditions, and practices such as rotating pools of savings, advice, and help in setting up a particular business. Ethnic and class resources complement each other for entrepreneurial success. Yet there are also limitations of ethnic resources. Ethnic resources can restrict enterprises to the ethnic market and can even lead to cannibalistic competition.[22]

Both types of resources operate within the scope of opportunities offered by an urban market. Among those opportunities are some that arise from ethnic community structures. The market for a Bollywood movie store obviously is circumscribed by the population of South Asian communities. An Indian may have an edge in this business, but it may remain small.

Other opportunities arise from the mainstream market and institutions. They are embedded in the local and regional economy, with place and location playing a significant role. They require acculturation in the mainstream culture, common language, and intercultural dealings. This is where public programs for business development, tax incentives, training, and local economic development strategies come into play, facilitating ethnic integration into the mainstream. Forward-looking ethnic communities also foster some form of mutual learning and acculturation through business associations, cross-cultural chambers of commerce, nationality networks, and ethnic media. These practices lay the groundwork for the formation of ethnic strategies that connect opportunities with resources. All in all, ethnic entrepreneurship is based on a combination of social and cultural qualities internal to a group and the opportunities/constraints presented by the economic environment. The idea of mixed embeddedness discussed above captures this reality.

Ethnic economies are not static. They evolve in response to changing economic trends and shifting local conditions. The demographic and institutional context of a city has a direct influence on the structure of ethnic economies. Eric Fong and Linda Lee identify four factors originating from the city that affect the performance of ethnic economies, namely: (1) the size of an ethnic group; (2) the level of residential segregation; (3) ethnic employment rates; and (4) the overall employment rate in the city.[23] How cities influence the growth of ethnic economies is a critical question from the perspective of multicultural cities. A complementary question is, what are the standings of different ethno-racial groups in incomes and employment (i.e., outcomes of ethnic economies and occupational clustering)? The rest of this chapter will address these questions.

The Urban Economies of the Three Cities –
Opportunity Structures

The Toronto Census Metropolitan Area (CMA) is the smallest of the three metropolises. Yet it is both branded and regarded as a paradigmatic multicultural city. Its population of 5.8 million (2011) makes it the largest urban area in Canada, and the inflow of about 80,000 immigrants per year is the lifeline of its population and labour-force growth. Its economic base is diverse. Trade (16.6%), education and health services (14.4%), and manufacturing (13.5%) had the largest share of the employed labour force (table 5.2). Manufacturing has remained robust, and Toronto's role as the financial centre of Canada and the premier city for health and educational services give depth to its economy. It has gone through cycles of expansion and recession since the 1990s; the recovery from the last recession of 2008–9 has been slow but steady. Its structure of opportunities has led to the formation of ethnic niches of Chinese, Italians, and South Asians in trade, fashion and garments, construction, trucking and transport, and computer technology.

New York and Los Angeles are the two biggest global cities of the United States. New York's population of 8.18 million (2010) makes it the largest regional economy in the States. Services altogether employed 76.7% of employed labour in 2009. Health and education (24.8%), trade (13.5%), and financial services (13.4%) formed the economic base of the city (table 5.2). It has the United Nations and draws large numbers of tourists, visitors, and officials from abroad. New York is home to creative industries, mass media, the arts and theatre. It has a declining but still robust manufacturing sector making garments, processed food, metal products, and software. The city has the headquarters of forty-three Fortune 500 corporations. There is a big market for ethnic businesses serving their communities, other New Yorkers, and visitors. Immigrants and minorities form niches in mainstream jobs and businesses and spawn ethnic economies. Taxi services are often the entry niche for newly arriving immigrants. The construction industry is dominated by Italians. West Indian nurses form a niche in medical services, as do Jewish doctors. The city's economy swings with the business cycles of the national economy. It has recovered steadily from the recession of 2008–9. Catastrophic events such as the terrorists attack in 2001 and "Superstorm" Sandy in 2012 shocked the local economy, but it bounced back quickly.

Table 5.2 Percentage of labour by industry

Industry	Toronto CMA (2006)*	New York City (2009)*	Los Angeles County (2009)*
Manufacturing	13.5	2.4	11.3
Construction	5.4	3.7	3.5
Trade, wholesale and retail	16.6	13.5	17.6
Finance, insurance, real estate	9.4	13.4	6.8
Education, health	14.4	24.8	16.1
Entertainment, accommodation, arts, and food services	7.7	10.3	11.4

Sources: For Toronto – Statistics Canada, Census 2006, Table, " Profile of Labour Market Activity, Occupation, Education for Census Metropolitan Areas and Agglomerations." For New York and Los Angeles – David Gladstone and Susan Fainstein, "The New York and Los Angeles Economies from Boom to Crisis," in *New York and Los Angeles*, ed. David Halle and Andrew Beveridge (New York: Oxford UP), 89, table 3.7. Based on census data.

*Note: The census organizations of both Canada and the United States have changed the data collection practices for socio-economic variables from the long-form censuses to sample surveys. The data for New York and Los Angeles came from annual American Community Surveys (ACS). Toronto's data came from the last long-form census, of 2006, after which Statistics Canada shifted to an optional decennial sample of the National Household Survey, 2011, which is not yet comparable to the long-form census in reliability.

Los Angeles County, including the city, had a population of 9.82 million in 2010. Its demography is different from New York's. So is its economy. Los Angeles County's economy has lesser concentration in services (63.0% of employment in 2009) than New York, but manufacturing had a higher share (11.3% in 2009), inherited from defence industries established during the Second World War. Table 5.2 shows that trade (17.6%), education and health, services (16.1%), and entertainment (Hollywood) and food (11.4%) were the leading industries in 2009. Los Angeles has an industrial cluster of information, defence, biomedical, and environmental technologies, which is an emerging driver of its economy and a platform for international trade. It is the biggest American port and is the connecting point for the Pacific trade. This is the framework of opportunities within which ethnic entrepreneurship and economic participation operate.

Photo 5.1 Chinese shopping arcade inside Pacific Mall, Markham (Toronto) (courtesy Susan Qadeer)

The Latinization of Los Angeles is the storyline of its transformation under the influence of immigration. Miles Davis observes that "Latinos have become predominant in low-tech manufacturing, home construction and tourist-leisure services," establishing ethnic economic niches and enclaves.[24] Chinese and Indians are filling the ranks of professionals in the new knowledge economy, whereas Chinese as well as Koreans and Japanese have helped forge cross-Pacific trade through their transnational links.

The three cities have many common elements, namely, (1) large and diverse populations, (2) being biggest cities of their countries, (3) economic bases in financial, educational, health, and producer services; and (4) global connections and a role as pre-eminent information, entertainment, and art nodes. Their differences arise from the scope and scale of their local and national economies. Structurally, their economic bases are rooted in slightly different mixes of industries, including manufacturing. The ethno-racial mix of their populations is also different.

Ethnic Entrepreneurship and the Formation of Economic Niches in Toronto, New York, and Los Angeles

The employment and business opportunities in the three cities are nominally open for all eligible persons equally, but in practice their labour markets are divided into segments by professional and occupational requirements, informational costs, and locational and accessibility factors. Social barriers further divide these market segments along the lines of racial and ethnic preferences, cultural and language predispositions, and most of all by what employment counsellors commonly call networking practices and personal connections. The segmentation of opportunities lays the groundwork for the formation of ethnic niches and clustering by ethno-racial groups in some activities and sectors. In particular, immigrants facing downgrading of their credentials and experiences fall back on their entrepreneurial talent and gravitate towards self-employment in ethnic economies.

Place matters in the access to economic opportunities. Apart from the opportunity structure of a city, its civic culture and public infrastructure and policies have influence over the channelling of access. Specific entrepreneurial idioms of ethnic businesses emerge in particular cities. Ethnic restaurants, craft shops, or grocery stores are such idioms that, though common across cities, are honed locally. They become "formula" businesses; the knowledge about local sources of finance, supplies, labour, and locations is passed around within an ethnic group.[25] To start such enterprises becomes a part of the entrepreneurial repertoire of a community. An example of such a formula business is the dollar store that has become a niche for Pakistanis in the Toronto area. It involves low investment and the information and contacts are readily available in the community. A new entrepreneur has a laid-out path to follow. How do these factors lay down the ethno-racial contours in urban economies of the three cities? This question is examined in the following sections.

Toronto

Ethnic economies in the Toronto area are primarily in the form of enclaves based in ethnic neighbourhoods. Space and location play a significant role in the establishment of the ethnic economic enclaves of malls and plazas (discussed in chapter 4). Brampton, a suburban municipality, is now dominated by South Asian, particularly Sikh, businesses and professionals. Thanks to its proximity to the airport, it is also the home of trucking services, which are another Sikh niche. Similarly, Markham and Richmond Hill, two other suburban towns, are the sites of primarily Chinese, but also Iranian and Jewish, business clusters. Restaurants, banquet halls, jewellery and gold shops, as well as apparel and household goods stores clustered in strip malls of ethnic identities line the main roads of these towns. Ethnic businesses have also made inroads into the mainstream economy. The Chinese have a niche in computer hardware. Italians and South Asians dominate construction.

In the city of Toronto, financial, real estate, educational, and health services have been dominated by native-born of European ancestry and Jews, but now young professionals of Asian backgrounds are making inroads into these professions, forming informal networks and nationality-based professional organizations. Filipino nurses and nannies are an economic niche by themselves. Almost all major ethnic groups now have evolved networks of ethnic businesses and professionals large enough to merit their respective business directories and run commercials on multicultural TV and radio channels, pushing their advantage as those who speak "your" language.

One common indicator of entrepreneurship is self-employment. It is a widely used indicator as the references to self-employment studies below will show. Self-employment as an indicator of entrepreneurship includes both small-business owners, such as ethnic taxi owners and restaurateurs, as well as the owners of corporate businesses in the mainstream. As most ethnic businesses begin small, self-employment tends to capture small entrepreneurs, which reflects the empirical reality of ethnic entrepreneurship.

Michael Ornstein of the Institute of Social Research, York University, in Toronto has analysed data on self-employment by ethnicity from the 2001 census. His report shows that the self-employment rate for the total male employed labour force (18–64 age groups) in the Toronto CMA was 14.2%.[26] Different ethnic groups show wide variations in the rates of self-employment. Among those with rates of self-employment significantly higher than the CMA averages are Koreans (30.9%), Jews (29.3%), Taiwanese (27.5%), Armenians (26.1%), Greeks (21.1%), Americans (20.7%), and Italians (16.9%). Chinese (14.9%), Pakistanis (14.1%), and Punjabis/Sikhs (14.1%) were close to the area average, while Jamaicans (7.7%), Mexicans (9.5%), Vietnamese (4.5%), and Filipinos (3.5%) had a self-employment rate significantly below the area norm.[27] Women's self-employment rates were lower than men's, but ethnic differences mirrored the men's profile.[28] A noteworthy fact of these figures is that almost all groups of high self-employment rates were relatively small communities of ethnics differentiated by their nationalities, either presumably close knit or those with niches in professions (e.g., Jews and Americans). Some large communities are known to have cultural resources predisposing them for businesses (e.g., Italians, Greeks, Armenians). Those with low rates of self-employment were generally groups whose skills and traditions predisposed them for other types of work.

New York

The city of New York fully reflects its majority-minority characteristics in its economy. Immigrants were 45% of the labour force despite being 36% of population in 2009.[29] The city has a bipolar job market, where jobs are concentrated at the high end in finance and business and the low end in the service sector.[30] Ethno-racial minorities, particularly immigrants, are represented at all levels, but they are clustered in the shrinking middle. US-born non-Hispanic Whites and Blacks have declining working-age populations, and the looming labour shortfall is being filled by immigrants.

Ethnic economies in New York are relatively small and shifting. The Chinese economy is the largest and most enduring. I will discuss it in detail later. Ethnic/immigrant economic niches are more common, but they also shift over time. About half of all immigrants work in white-collar jobs, as do 75% of the native-born, including ethno-racial minorities. Despite New York's openness, race and ethnicity are predictors of how workers will fare in the economy at all levels.[31] There is a premium in incomes for US-born Whites for all levels of education, followed by US-born Asians, Blacks, and Latinos. Immigrants of equal education earned 10%–20% less than their US-born counterparts of the same race and ethnicity.[32]

Ethnic clusters are discernible all across the employment spectrum. Indians, Jews, Russians, and Hong Kong–born Chinese are concentrated in financial, managerial, and professional occupations. Whereas Mexicans and mainland Chinese were concentrated in food services, Russians, Jews, and Indians had a substantial presence in professions and managerial jobs. Native Blacks were largely clustered in public-service jobs, while West Indian women were concentrated in home care and nursing. Anglo-Saxons had ethnic niches in law, medicine, banking, and corporate management. Hispanics and South Asians are forming niches in home repairs and small-scale construction. Taxi services have been the ladder for new immigrants to enter the city's economy. Russians, Jews, Israelis, Pakistanis, and now Bangladeshis have successively circulated through the taxi services business, forming and dissolving ethnic niches.

The story in self-employment is not much different. Greeks (29%), Koreans (26%), Italians (19%), Pakistanis (18%), and Russians including Jews (16%) had self-employment rates significantly higher than the average for the city's labour force (11%). Among women workers, Koreans (15%), Greeks (13%), Colombians (13%), and Mexicans (11%) scored much above the city's average (6%) in 2000.[33] Koreans were largely concentrated in corner grocery stores, laundries, dry cleaning, and importing of goods from Korea. Their limited facility in English and inability to practise their professions drove them to lower-circuit entrepreneurship. Italians and Greeks were in the construction and maintenance industries. Mexicans and Colombians were concentrated in restaurants, groceries, construction, and domestic services; many were self-employed in lower-circuit enterprises, some even as day labourers. These ethnic niches are the outcomes of the respective ethnic groups' strategies to leverage their human and cultural resources in the competitive New York job market.

The city's economic structure itself has been buffeted by global and national trade and economic policies. The decline of manufacturing in the United States and the erosion of steady unionized industrial jobs have increased the city labour force's dependence on contractual jobs and self-employment. Immigrants have little choice but to be enterprising, and are increasingly (particularly new arrivals) dependent on their community networks for gainful employment. Therefore, network-driven employment fosters ethnic niches and enclaves.

Los Angeles

The Los Angeles metropolitan area's economy was restructured in the 1990s by the recession, reduced defence spending, and globalization,[34] which has made self-employment increasingly necessary for the jobless and provided a stimulus for creative enterprises of the new economy for investors and professionals. On the lower end, casual workers, sweat-shop labourers, and back-of-truck service contractors are expanding the ranks of self-employed entrepreneurs. In the middle and upper tiers of the economy, professionals are operating as independent producers and contractors. Self-employment is rising and ethnic niches are being carved.

In 2008, the percentage of unincorporated self-employed in Los Angeles County's labour force was 9.5%, much higher than the average for the United States (6.6%) and more than California's (8.8%).[35] Asian businesses were bigger and had almost three times the average sales per business compared with Hispanics enterprises and four times that of Blacks. Most of the businesses were individual/family enterprises running mom and pop stores or selling their labour. They represent the entrepreneurship of immigrants creating their employment.

As in New York, Koreans had the highest rates of self-employment (35%), followed by Iranians (28%), Armenians (27%), Russians (23%), and Chinese (17%), while Mexicans (7%), Filipinos (5%), and Blacks (6%) had self-employment rates less than the average for the area.[36] An interesting fact is that the self-employment rates of native-born (usually second/third-generation) ethnics were lower than those of immigrants. It indicates that entrepreneurship of the lower circuit in particular was an ethnic strategy necessitated by blocked employment opportunities for immigrants, low human capital and class resources, lack of knowledge of English, and structural conditions of the local economy, not excluding racism. Over time, immigrants become ethnic citizens, their access to jobs improves, and they move out of small businesses.

Hispanics, particularly Mexicans and native-born Chicanos, had occupational niches in manufacturing, construction, and gardening, but over time they have shifted to being heavy-truck drivers, equipment operators, and mail carriers, and so on.[37] Chinese, Japanese, Filipinos, and Indians, lumped as Asians, have been called a "model minority." They have high levels of education and their children do well in school, filling up colleges and graduating into professional job niches. The Chinese in particular also start businesses in engineering, computers, and software, blending their class and ethnic resources, including trade and financial links with China, Taiwan, and other countries of the Far East. There is a pattern of ethnic groups forming a network/niche in particular industries, somewhat in line with their human and cultural resources and opportunities.

Black Americans in Los Angeles present a polarized outcome of their performance in the labour market. While the income gap has steadily closed in the professions, low-skilled Blacks, previously employed in manufacturing, have lost out owing to the deindustrialization of producer-goods industries. The new low-skill industries, small businesses forming immigrant niches, leave limited opportunities for poorly educated Blacks.

Black women have moved almost completely out of domestic service into low-level clerical jobs, working in both the public and private sectors. Blacks also continue to have a niche in low- and mid-level public service jobs. One segment of the Black labour force has remained blocked, while another segment has gained and moved into new niches.[38]

The Convergence of Ethnic Strategies in the Three Cities

Ethnic entrepreneurial profiles, reflected in self-employment rates, are similar in these three cities, where narratives about cultural diversity vary considerably. Koreans, Jews, Greeks, and Chinese/Taiwanese were significantly more entrepreneurial than Hispanics, Vietnamese, Filipinas, and Blacks in the three cities. There were small differences in rates of self-employment for Pakistanis, Iranians, Italians, and Russians. They may stand out in one city but not in the others and vice versa, suggesting that local opportunities for their talent may have varied according to the size of their communities. Even the industrial concentrations of these ethnic groups were more or less similar.

Jews were concentrated in professions such as medicine, law, and the production of art, film, and theatre, forming ethnic niches in Toronto, New York, and Los Angeles. Koreans were greengrocers and gift-shop owners in neighbourhoods in all three cities. Greeks were restaurateurs and Italians had a niche in construction. A large number of Chinese were operating restaurants, groceries, and other small businesses, while another stream was at the forefront of information technology and computer design, and other niches in the mainstream. Sikhs were a prominent group in Toronto, and their niche in airport taxi service, trucking and transport, and repairs was unique to Toronto. Indians, wherever prominent, were concentrated in professions and new technologies. Blacks everywhere lagged in businesses and professional jobs. Similarity of ethnic niches in the three cities points to a parallelism of ethnic strategies and cultural resources among the respective groups.

Just as culture and race matters, so does place. New York has been consistently getting more polarized in income and employment, with the middle class shrinking and lower and upper classes expanding through all phases of the business cycles of the 1990s and 2000s. Los Angeles has also experienced polarization, but its disparities have steadied in the 2000s, perhaps with its increasing role as the connecting point for the Pacific trade, particularly with China, and unionization drives for service workers.[39] Manufacturing is still strong in Los Angeles.

Toronto is a smaller economy. It has retained manufacturing. The Canadian social-welfare state provides a floor, particularly through its public health insurance. The city's economic polarization has widened and shrunk with expansions and recessions, though in the past thirty years polarization has increased consistently, as has been shown by David Hulchanski.[40] These similarities and differences are reflected in the opportunity structures for ethnic economies and the economic prospects of immigrants.

Earnings by Ethnicity

Ethnics and immigrants are gainfully employed in three types of economic organizations. One, they work in the mainstream economy, integrated fully as employees and employers. Two, they work in the mainstream economy, but from the platforms of ethnic niches. Three, they work in ethnic economies as business owners and employees, which are embedded in the mainstream economy. Overall, a majority of ethnics and immigrants work in the mainstream economy as a part of the national labour pool.

Evidence about the percentage of the ethnic labour force working in ethnic economies is not only scanty but also contradictory. Ivan Light and Steven Gold conclude from various observations that in the United States about 41% of minority ethnic groups including immigrants, work in ethnic-ownership and ethnic-controlled economies.[41] Of course, there is a wide variation in the size of ethnic economies of different groups in various cities, and consequently in the percentage of the labour force employed in them.

Ethnic businesses are largely limited in scope, essentially providing earnings to the self-employed entrepreneur and creating relatively few jobs for employees. The data from the above reported study show that from 72% to 92% of Asian, Hispanic, and Black businesses in Los Angeles were purely family enterprises with no paid employees. Only the incorporated enterprises, presumably of the upper circuit, created extra jobs and paid handsome earnings for their owners.

A study of ethno-racial groups' comparative earnings in Los Angeles from 1970–90 shows that earnings from self-employment in incorporated businesses were the highest for each of the four ethno-racial groups, namely, Whites, Blacks, Hispanics, and Asians.[42] Even self-employment in unincorporated businesses provided higher incomes than employment in either government or private enterprises for three of the four groups.[43] Hispanics in government services were the only exception, earning more than from self-employment. Government jobs yielded higher earnings than employment in private organizations for three of the four groups; Whites in the private sector earned more than those in government jobs.

The above study of the Los Angeles labour force by ethnicity used regression analysis to control the effects of education, age, and English proficiency. It found that self-employment provided higher incomes even when human capital was controlled for most major ethnic minorities. There were extra earnings in self-employment compared with the incomes of employed compatriots of similar human capital. A Chinese native-born (NB) self-employed person had a bonus of $14,232, the foreign-born (FB) of $12,993, Korean FB self-employed earned a premium of $9941, and Black NB self-employed earned $13,467 more than employees in public or private organizations.[44] The self-employed not only earn more than those who work as employees, but they also build up assets that further increase their lifetime earnings. There are some dissenting voices to this conclusion, but the balance of evidence supports this observation.[45]

Apart from education, age, skills, and experience, that is, the elements of human capital, cultural and social capital, nativity, race, and ethnicity affect earnings. It is therefore difficult to isolate the effect of ethnic or racial factors on earnings, particularly at the city level. There are two separate studies relating to two of the cities that concern us here. They give some consistent observations about earnings by ethnicity.

For the Toronto area in 2001, Michael Ornstein has estimated the differences in earnings of various ethnic groups after accounting for the effects of age, education, and immigration with a regression analysis. His overall conclusion is that after accounting for the three variables, non-European men had 20% less income than Europeans.[46] The gap is smaller for women.

Ornstein found that men of certain ethnic groups earned more than the average "Canadian," such as Japanese (110.2%), Egyptians (101.7%), and Lebanese (101.1%). Most others earned less, such as Iranians (83.2%), Indians (85.4%), Chinese (85.7%), Jamaicans (83.6%), Aboriginals (81.7%), and Blacks (75.0%). Yet persons of Northern European backgrounds, such as the English (109.8%), Americans (123.3%), Germans (109.3%), and Jews (130.2%), had incomes significantly above the Canadian average.[47] Comparable incomes for ethnic women were proportionately higher than for men, but lower than the Canadian average for women; for example, Iranian women earned 99.6% of the Canadian norm, Chinese 99.8%, Indians 91.8%, Jamaicans 86.4%, Aboriginal women 82.4%, and Black women 81.4%.[48] The income gap was almost eliminated for persons of mixed non-White/White heritage. There was an income gap between the ethnics of European ancestry and those of Asian, African, and Latin American origins for both men and women. This divide in incomes for persons of similar education, age, and immigrant status also demarcates the income disparity between Whites and non-Whites. Jeffery Reitz has documented that this income disparity has stubbornly persisted since the 1970s.[49]

David Kallick reports similar results for New York by comparing median annual wages by race or ethnicity in 2009. First, the US-born in all categories of education and ethnicity earned more than the foreign-born. Holding education constant, the US-born always earned more on average than the foreign-born for all racial and ethnic groups. Among US-born with high school education, earnings were, for Blacks 78%, Latinos 82%, and Asians 87% of Whites' average. Among college-educated foreign-born, earnings were Whites 88%, Blacks 71%, Latinos 58%, and Asians 74% of the US-born Whites' average earnings.[50]

What is obvious is that the earnings of non-White minorities, which included immigrants, are lower than those of the Whites of Northern

European extraction, even when the effects of education, age, and other factors of human capital are controlled. This pattern holds true in all three cities and in both the United States and Canada. There are also consistent differences of incomes among ethno-racial minorities. For example, Chinese, Iranians, and Indians of similar qualifications were earning more than Caribbeans, Hispanics, and, most strikingly, Blacks. Alejandro Portes and Ruben G. Rumbaut estimated the differences of income among the children of immigrants, who were college-educated, fluent in English, with twenty years' residence in the United States. They found that East Asians (viz., Cambodian/Laotian and Vietnamese) would have earned about $1000–$1500 per month more than those of Central American/Caribbean origin (Mexicans, Nicaraguans, and Haitians) in 1995 dollars.[51] They attribute these differences to what they call a "context of reception," including factors such as government policies, conditions of the host labour market (place effect), and the characteristics of an ethnic community.[52]

The White/non-White differences in incomes for persons of similar qualifications are often commonly explained as the result of entrenched racism. Undoubtedly, race discrimination has been a structural condition in both the United States and Canada. It must also be recognized that anti-racism measures have been a part of civil rights in both countries for almost forty years, and glaring incidents of racism in hiring and promotion get a lot of public attention. Yet the much-lauded networking in the job market erodes the equity of employment opportunities and favours the members of the dominant groups. This is one mechanism by which ethno-racial differences in income are systematically sustained; the rich and influential network with the rich and powerful and the poor network with their own kind. The economic and political power of an ethnic community affects the destinies of its members. Also, there is some premium on nativity as well as immersion in the mainstream culture.

The Multiculturalism of Consumers

Ethno-racial contours are mostly viewed from the production or supply side of the economy. Most of the literature and empirical studies are focused on the concentration of ethnic and racial groups in specific activities, employment niches, and businesses. But there is also the demand or consumption side of ethno-racial diversity. In particular ethnic and cultural differences in food, clothes, housing, leisure, and so on delineate distinct consumer submarkets. This culturally differentiated consumption results in the segmentation of an urban economy. It stimulates imports and spawns culturally and linguistically diverse services, such as marriage bureaux, funeral homes, physicians or dentists speaking relevant languages, and Halal or Kosher butchers.

In the three cities, there are well-structured ethnic networks of the suppliers of goods and services. Such ethnic networks have their respective business directories, websites, radio and TV commercials, and advertisements in ethnic newspapers that facilitate consumers' access to all kinds of goods and services, both ethnic and mainstream. Chinese, South Asians, Latinos, Italians, Russians, Iranians, Arabs, and immigrant Africans have extensive ethnic networks of suppliers for consumer demand in the three cities. It is possible to live, for example, within an ethnic bubble of construction contractors, physicians, nannies and caregivers, investment advisers, homeland TV and music concerts, as well as ethnic food and clothing stores, for meeting everyday needs. These markets are open to others and they operate within the legal and financial institutions of the mainstream.

Given the demographic reality of an increasing ethnic population and relative shrinking of the number of Anglo-European consumers, many mainstream supermarkets and stores are stocking ethnic goods. There are Halal meat shelves, racks of spices, and frozen foods of different nationalities in the supermarket chains of the three cities. Ethnic products are being integrated into mainstream commerce. Recently, corporate telemarketers have started using ethnic languages and salutations to target ethnic consumers. For example, telephone and Internet service providers in Toronto use ethnic speakers to promote their offerings. The same is true in New York and Los Angeles. All in all, consumer markets of the three cities are segmented along ethnic lines to some extent, though open to outside influences.

There is some indication that ethnic consumers show a preference for ethnic stores and service suppliers if those are readily accessible and competitive in price. A survey of middle-class Chinese consumers in Toronto found that they showed a strong preference for Chinese travel agents and a tendency to divide their grocery shopping between the ethnic and mainstream supermarkets.[53] Chinese are not alone in showing a preference for ethnic suppliers of goods that have strong ethnic identities. Other ethnic groups show similar tendencies. Ethnic consumers' preferences help sustain ethnic economies.

Reprise: Ethnic Alignments in the Urban Economy

Ethno-racial diversity realigns the economic organizations of multicultural cities. It leads to the segmentation by ethnicity and race of both the job and consumer markets. This segmentation does not wall off various groups from each other. It is fluid, with movement from ethnic economies to mainstream employment and businesses. But the concentration of different groups in particular activities, occupations, and social statuses is evident in the three cities. They have well-structured subeconomies, employment niches, and networks of ethno-racial identities. Immigrants particularly converge on such cultural-economic clusters to leverage their cultural resources, connections, and social capital, over and above their human capital, into employment and income. Continuing immigration is likely to make ethnic economies and network an enduring feature of urban economies. The following are observations drawn from the foregoing analysis that sum up the ethnic alignments of the three cities' economic organizations.

1. By and large, the entrepreneurial strategies of an ethnic group are similar in the three cities of the United States and Canada. These two countries have distinct patterns of multiculturalism, yet ethnic entrepreneurship takes similar forms.
2. Ethnic entrepreneurship opens the path to economic advancement, particularly in the upper echelons of the local economy. In the lower-circuit activities, it inhibits immigrants' integration into the mainstream, but provides a way around their blocked opportunities.
3. Social networks and community-based connections and trust help forge ethnic economies, taking advantage of particular ethnic groups' knowledge and talent for some activities and occupations. Many ethnic businesses become "formula" activities for which information, connections, and resources are shared among co-ethnics. That lays the path for ethnic groups to concentrate in particular activities and professions.
4. Ethnic minorities and immigrants have higher rates of self-employment than the city averages in all three cities. Some ethnic/national groups have much higher rates of self-employment than others in these cities, for instance, Koreans, Italians, Jews, Greeks, and Chinese/Taiwanese. Others do well in New York, but not in Los Angeles or Toronto, and vice versa, such as Russians, Pakistanis, and Armenians.

5. Among ethnics and immigrants, rates of self-employment and levels of income vary from group to group. Koreans, Chinese, Indians, Iranians, and in some cases Vietnamese/Cambodians have higher rates of self-employment and earn more than Mexicans, Nicaraguans, and Jamaicans.
6. Non-Whites in general earn less than Whites even when their talents and resources are comparable. Also, there is an earning premium to being native-born in both the United States and Canada for persons of similar backgrounds.
7. Average earnings of Whites are the highest, followed by Asians, Latinos, and native-born Blacks. Blacks and Latinos lag behind in earnings even when human-resource factors are accounted for. Is this a legacy of racial discrimination?
8. Generally, incomes from self-employment are higher than from salaried jobs for persons of comparable talents and similar ethnocultural backgrounds.
9. Ethnic segmentation also operates in consumer markets. For some groups it is possible to meet most of their everyday needs within their ethnic network.
10. Ethnic economies are fully embedded in urban economies. They are subject to the same labour, finance, public health, and taxation regulations, though accommodations are made to respond to cultural/religious differences. Similarly, they are subject to national and local business cycles, global trade patterns, and technological change. Yet the common ground of local and national economic order frames the opportunities and structures of ethnic economies and niches.
11. Ethnic economies are a relatively small part of urban economies. A majority of ethnics and immigrants work in mainstream establishments. Yet their entry into the labour market is facilitated by formal and informal social networks, which tend to be based on ethnic ties.

Urban economies are not undifferentiated economic organizations. There are distinct patterns to the clustering of economic activities and earnings by race and ethnicity. These are the structural features of the urban economies of multicultural cities, as observed in Toronto, New York, and Los Angeles. Are there local differences in these aspects? Although this question has been probed above, an opportunity for a social experiment presents itself in the form of the comparison of Chinese economies in the three cities. Holding ethnicity constant and observing variations of the cities' economic opportunities present an intriguing situation to observe ethnic economic structures and processes. There are extensive studies of the Chinese economies of the three cities on which I have drawn for analysis.

Similarities and Differences of the Chinese Economies in Toronto, New York, and Los Angeles

These three cities historically have been the premier destinations for immigrants in Canada and the United States. They are multi-ethnic, multicultural, and either are, or are at the cusp of being, majority-minority places. The Chinese in each city have a long presence and the trajectories of their Chinatown-initiated economies have many parallels.

The Chinese were among the early non-White immigrants in the United States and Canada, having started to migrate to both countries to build railroads in the 1850s. Strikingly, the reception of the Chinese in both countries was similar: allowing only male labourers, barring their families, and then enacting the exclusion acts ending Chinese immigration altogether. Eventually, after 70–90 years, the acts were repealed, in Canada (1947) and in the United States (1943). Chinese immigration began in earnest after 1965 with the removal of country quotas and the introduction of occupational preferences and family entitlements as the basis of immigration to both countries. It has picked up pace over the years and, more importantly, the composition of Chinese immigrants has changed. Not only workers from Fujian and Guangdong provinces, but also professionals and investors from Taiwan, Hong Kong, and the mainland are migrating to the United States and Canada. With the rise of China and booming Western trade and investment relations with Chinese, the immigrants' prestige and acceptability has risen. They are branded as a "model minority" and their professional and academic achievements in the United States and Canada have earned them envy. The historic racism remains a faded hurt.

The Chinese now form sizeable communities in each of the three cities. The single-heritage Chinese population was 531,635 in the Toronto CMA (9.6%) in 2011, 487,532 (5.9%) in New York in 2010, and 402,562 (4.1%) in Los Angeles County in 2010.[54] Chinese of mixed race/ ethnicity were in addition. These populations have grown largely in the past three decades. For example, 64.4% of New York's Chinese came after 1990.[55] The new wave of immigrants has promoted the growth of

Chinese economies and niches in the three cities. A three-phase account of the process is presented here to help clarify the dynamics of Chinese economies in the three cities.

Phase 1: The Formation of Chinatowns and Ghetto
Economies (1850s–1950s)

This phase lasted from the early settlement of Chinese immigrants to the end of the exclusion acts. In all three cities, early Chinese migrants huddled together in a few blocks of the downtowns. Their grocery stores, eateries, and legal and community services formed the nuclei of Chinatowns. Chinese restaurants and laundries were the activities that connected Chinatown to the rest of the city. A 1923 study of Chinatown in Toronto enumerated 203 restaurants, 47 laundries, and 7 groceries.[56]

The popularity of Chinese food among cosmopolitan New Yorkers made its Chinatown an eating-out venue. The budding Chinese economy was centred in Chinatown, whence even the scattered laundries in the city were managed in various ways. The finances were supplied by rotating credit associations, and workers as well as owners converged on Chinatown for services, companionship, and entertainment.

Los Angeles's Chinatown began where Union Station, the railroad terminal, stands today, but its demolition spurred the development of a New Chinatown in 1938 a few blocks away. It was a small cluster of Chinese restaurants, groceries, and curio shops serving the Chinese population. It differed from Chinatowns in the other two cities in that it served as a ready-made set for Hollywood's romantic Orient.[57] Thus, it paralleled the Chinatowns of Toronto and New York in its origins, though with a Hollywood twist.

By the 1930s and particularly during the Second World War, the Chinese were viewed as allies. Their numbers steadily increased and Chinese residences and businesses spilled out of the historic Chinatowns. This happened in parallel in all three cities. Richard Thompson calls it "a transitional phase" for the Chinese economy in Toronto.[58] A 1966 survey revealed 448 Chinese firms in Toronto with proportionately fewer restaurants, more laundries and dry cleaners, and new types of businesses such as import and export, gift shops, real estate, insurance, travel agents, and professional services.[59]

Centred on Chinatown, Chinese economies had remarkably parallel developments in Toronto, New York, and Los Angeles. Their history as ghetto commercial strips was parlayed into being tourists' attractions.

Yet there were differences arising from the dynamics of local economies and consumer tastes. Toronto's Chinatown remained largely an immigrants' neighbourhood with restaurants and laundries. New York's proclivity for entertainment made its Chinatown restaurants and import emporiums city-wide businesses.

Phase 2: The Diversification and Expansion of Economic Activities (1960–1990)

The opening of immigration to professionals, investors, and families of immigrants after the rescinding of country quotas in both countries in 1965 started the new wave of large-scale migration to Canada and the United States. International events, such as the Vietnam War and the subsequent refugee crisis, Hong Kong's return to Chinese sovereignty, combined with the emigration of Chinese who were "middlemen minorities" in Southeast Asia, increased the tempo of Chinese immigration. US recognition of communist China (1979) opened the door to Chinese from the mainland. All these events brought new classes of Chinese immigrants to the gateway cities of Toronto, New York, and Los Angeles. They were professionals, entrepreneurs, and businessmen who joined workers from the traditional sources of Chinese immigration.

In the same period, the economies of these cities also underwent extensive restructuring with the introduction of information technology, deindustrialization, the flourishing of producer and social services, and the opening of international trade and financial flows. The new talent pool of immigrants, combined with the changing opportunity structure of urban economies, realigned the Chinese economies in the three cities.

The Chinese economy in the Toronto area grew new shoots, diversifying its sectoral composition and establishing new niches in the mainstream economy, for example in finances, real estate, fashion, and computer engineering. A new idiom of the commercial centre in the form of Chinese or Asian malls was forged. It is the suburban counterpart of Chinatown. Its innovation lies not only in its architecture and organization but also in the condominium form of store ownership, which expanded the scope of investment for small investors. About twelve such malls and another forty-three plazas of three or more Chinese stores were built in this period.[60] The Chinese economy in Toronto has been anchored in Chinese territorial enclaves, centred on suburban malls and neighbourhoods.

In New York, a garment-industry nucleus had formed in Chinatown by the 1960s, complementing the wholesale trade, retail businesses, and financial and service enterprises that had emerged over time.[61] The finance, insurance, and real estate and producer services, including Chinese banks, were the new activities added to the Chinese economy. New Chinatowns in Flushing and Sunset Park emerged as the focal points of new immigrants differentiated by their countries of origins. The Old Chinatown has expanded into contiguous parts of lower Manhattan and has become an overflowing centre of Chinese imports and knockoff designer T-shirts, watches, shoes, and other personal accessories since the 1990s. It also supports a jewellery niche, specializing in low-price gold and diamond pieces. Yet it retains its function as the place to go for Chinese meals.

In Los Angeles, the Chinese economy found niches in durable goods manufacturing in the 1960s and then growing trade with Japan, China, and other Pacific Rim countries in the 1970s and 1980s. A robust Chinese banking sector, with branches of mainland and Taiwanese banks, emerged that has funnelled Chinese investments into finance, insurance, and real estate as well as imports and exports.

The centre of gravity of Los Angeles's Chinese economy shifted to the emerging ethnoburbs, spawning economic enclaves differentiated by social identities of Chinese provenance, such as Little Taipei in Monterey Park, a high-status enclave in San Marino, and a middle-class mini-city in Alhambra in the San Gabriel Valley.[62] A social sorting process has swept through both native-born Chinese as well as immigrants, clustering them by their origin, dialect, and social standing. Mandarin-speaking Taiwanese head to Monterey Park drawn by the real estate advertisements in Taiwan to potential immigrants and investors,[63] whereas rich Hong Kong and mainland Chinese converge in San Marino and the central city, respectively. Min Zhou et al. describe San Gabriel Valley's Chinese economy as an economic enclave that has occurred with Taiwanese investments in real estate, which developed commercial centres, business parks, clubs and entertainment complexes, and housing.[64]

Phase 3: Global and New Industrial Developments (1990–)

In this phase, the trends emerging in phase 2 have continued to rise and broaden. A major new force is the globalization of national trade and production. Chinese ethnic economies and niches have served as the conduit for global trade, production, and distribution in both Canada and the United States. Chinese investments in information technology and the clustering of Chinese professionals in computers and software industries, finance and banking, as well as real estate, show the maturation of Chinese economies and their embeddedness in the mainstream.

In the Toronto area, the Chinese economy has become a network of ethnic commercial enclaves and industrial niches. Lucia Lo and Shuguang Wang observe that the Chinese economy "has moved away from its traditional focus on consumer goods and services to a whole array of industrial and commercial activities, including producer and professional services."[65] Wholesale trade and manufacturing had become the largest component of the Chinese economy, with thirty-two firms employing more than one hundred persons.[66] The Chinese economy has developed an upper circuit with modern management practices and multi-branch operations. It is broadening into real estate development, funnelling Hong Kong and mainland Chinese investments into shopping malls, offices, and high-rise condos.

New York's Chinese economy, though dispersed over three enclaves, is essentially interconnected by forward and backward linkages among businesses and labour. Flushing has a big indoor mall and a business/professional centre. It brings Chinese, Korean, and Japanese businesses onto a common platform. The city has supported the development and modernization of garment shops as well as other Chinese businesses. The unionization of garment workers is supplemented by support for owners and improvement of the neighbourhoods' infrastructure and zoning.[67] New York's Chinese economy is based on relatively low value-added production. In the 2000s China's investments abroad are beginning to filter down to the Chinese economy in New York. Chinese banks and overseas investors are turning to real estate development, particularly in the gentrification of Manhattan's Chinatown.[68] New York as a tourist destination has contributed to the upgrading of Chinese restaurants and the revitalization of the historic Chinatown.

In the Los Angeles area, clusters of "industrial parks, warehouses, shopping centres and multifunctional office structures" have complemented the growth of Chinese residential suburbs.[69] These economic nodes are niches of ethnic business activities that are either being vacated by the larger economy or are new fields just opening up, particularly computer industries, high-tech manufacturing, or consulting services. The Chinese enclaves in the San Gabriel Valley have been described as "probably the single most important center" in the United

States for trade with China and the Far East.[70] In phase 3, Los Angeles's favourable location for the Pacific trade has been a transformative factor. The new high-tech manufacturing evolving in parallel with the historic defence industries of Los Angeles is drawing Chinese engineers and investors and incubating Chinese niches.

Chinese financial institutions have grown into an industry by themselves in the Los Angeles area in the 2000s. Los Angeles has forty-four banks operated both locally and internationally.[71] So significant is Chinese banking for the Los Angeles economy that most ethnic banks (about 20% of all bank branches) are Chinese of Hong Kong and Taiwanese origins.[72] New York and Toronto do not have as strong a Chinese banking sector as Los Angeles.

The Parallels and Divergences of Chinese Economies

The evolution of Chinese economies has followed a strikingly similar path in the three cities. They grew out of the neighbourhood commerce of early Chinatowns into the present ethnic niches of producing goods and services for metropolitan markets and global trade. Their economic organizations have evolved by channelling immigrants' talents and entrepreneurship, complemented by ethnic networks and cultural resources, towards opportunities arising in mainstream economies. As urban economies of the major North American cities have converged towards similar activities, such as services, finance, and information technologies, Chinese economies have tilted towards such activities.

Chinese economies, like the cities, have suburbanized. Chinese enclaves comprising malls, office complexes, and trade and manufacturing centres have emerged in the new Chinese neighbourhoods of the three cities. Toronto and New York are somewhat similar in that their suburban Chinese developments are mixtures of ethnic and mainstream economic activities that are linked with their historic Chinatowns. Los Angeles's Chinese economy has shifted largely to the ethnoburbs of the San Gabriel Valley, where ethnic establishments are backed by the social and political institutions of the Chinese communities.

Differences in the Chinese economies arise from the differing opportunities in the three cities, which in turn are the result of their economic and political roles in the national economies. Los Angeles's Chinese economy has been a bit different from those of New York and Toronto. It had a Hollywood connection even in its ghetto days. Its defence industries led to the high-tech-based manufacturing that drew on Taiwanese investments and professional engineers of Chinese origin.

Other differences reflective of the variations in the opportunity structures of the three cities are (1) in New York, a flourishing of the garment and jewellery industries and cheap-knock-offs trade employing largely Chinese immigrants drawn to the city for its long history of generous social services, (2) Toronto's Chinese niche in fashion and computer hardware sustained by immigrants from Hong Kong, (3) Los Angeles's banking and finance, dominated by overseas Chinese banks initially drawn by the Pacific trade, and (4) the recent emergence of Chinese real estate developers in the inner cities of Toronto and New York and in Los Angeles's suburban municipalities. All these differences arise from a combination of local economic conditions and the characteristics of immigrants drawn to a place.

Finally, the welcoming context for immigrants and responsiveness to ethno-racial diversity are the necessary conditions for the flourishing of Chinese economies. The right to equality in economic pursuits and freedom of cultural and linguistic expression, in sum, multiculturalism, are the contextual elements that help in promoting ethnic entrepreneurship and creativity.

The Social Benefits and Costs of Ethnic Economies

An ethnic economy is not a self-contained economic entity. It is embedded in the surrounding urban economy and polity. There is a continual flow of ideas, goods, services, finance, regulations, and labour between them. Regional and city infrastructure, planning and zoning by-laws, economic and business policies, licensing and health regulations have equal bearing on ethnic economies.

Communities of consumers and producers and the clustering of activities in ethnic economies precipitate external economies, thereby lowering costs, increasing efficiencies, and facilitating transactions. The social organization of ethnic economies binds together consumers and producers as well as owners and workers. Ethnic economies introduce new activities into the local economy, contributing to its growth and stability. The magic of multipliers works, raising incomes and expanding the economic base. As case studies of Chinese economies show, investment and talent are also drawn from abroad, stimulating economic growth and increasing the productive capacities of the local economy.

In the matter of demand and supply, ethnic economies thicken and broaden the mesh of trade and production and tap diverse resources and opportunities. In this era of globalization, they function as platforms for international investment and trade. As Florida has argued, cultural diversity is a source of creativity and innovation.[73]

Ethnic economies create jobs for immigrants who find themselves barred from the mainstream economy for lack of skills, language, or documentation. Ethnic economies function as segmented labour markets for ethnic and immigrant communities. They incubate new businesses and diversify the economic base. This diversification of activities has the effect of stabilizing local economies in times of cyclical swings.

The social costs of ethnic economies include displacing some business activity from existing enterprises and employees. Their segmentation of markets, while giving opportunities to immigrants, offers lower wages and limited scope to gain skills and knowledge, organizational and cultural in particular, for working in the mainstream. As Philip Kasinitz et al. observe: "Ethnic employment may well be preferable to unemployment, but it is a safety net, not a springboard."[74]

There is a general consensus that wages in ethnic economies, particularly in their lower circuit, are lower than in the mainstream economy, even if they are adjusted for productivity.[75] The self-employed fare better, as they build up assets resulting in two to fourteen times as much net worth as their salaried counterparts.[76] There are, of course, local variations. Among the self-employed entrepreneurs are business owners, physicians, and software designers, but this category also includes day labourers, temporary office workers, and street vendors.

The mortality rates for ethnic small businesses are high. A study of Chinese businesses in Richmond, a suburb of Vancouver, found that the survival rate of retail establishments after ten years was 41%.[77] Of course, mortality rates of small businesses in general are high, but ethnic economies largely based on small enterprises are particularly vulnerable.

Ethnic economies can be "gated economies" from which persons not belonging to a particular ethnic group may be excluded in terms of employment and investment. The exclusionary potential on the labour market can impede the openness of the local labour market and jeopardize employment equity. The common ground of the area-wide labour market has to be strengthened through public regulations and incentives. It is a social and moral imperative. Eric Fong and Emi Ooka observe that working in the ethnic economy hampers participation in the social activities of a wider society.[78] Ethnic economies operate in informal modes in their lower circuits, employing undocumented workers and carrying on many transactions off the books.

Taken altogether, a majority of ethnic workers fill the ranks in the labour force of the mainstream economy vacated by aging and retiring workers and by the expanding economy. Immigrants' entrepreneurship also grows by taking advantage of these opportunities. Ethnic economies are an important element of the local economic organization, but they thrive within the framework of local opportunities. They serve as the entry point for immigrant workers, leading to their eventual integration into the mainstream market of jobs and businesses.

The "New" Urban Economy and the Role of Ethnic Economies

Towards the end of the twentieth century, changes in national and global economic organizations have been restructuring urban economies. Information technology and free trade have combined to generate a techno-institutional storm that has changed the economic structure of cities in general but of North American cities in particular. Unionized manufacturing jobs, the staple of urban economies, have been shed massively and services are becoming the drivers of urban economies. But information technology is realigning the production of not only materials and goods but also professional and personal services.

The click of a key can send a letter, blueprints, manufacturing instructions, or money across the ocean in a second. Even professional services are being decentralized to distant sites; for instance, in some US hospitals, interpreting X-rays is now electronically outsourced to radiologists in India and Thailand.[79] Call centres in India are pitching rug-cleaning services to households in Toronto or New York. Such are the realignments of consumer markets. The rapid circulation of information and capital is stimulating the movement of workers across national borders. Tasks that engaged numerous workers previously are increasingly being done by a few with the help of computers. The privatization of public services is further breaking up stable jobs into contractual work. The opportunities for middle-class occupations are being squeezed out. Social disparity between the top and middle-bottom tiers is increasing.

A new global economic order has emerged in which some cities are the command centres of the global economy and others are satellites serving as the sites of secondary production. These forces are polarizing cities' job markets into circuits of high-paying professional and managerial occupations, on the one hand, and low-paid service and manufacturing jobs, on the other.[80] This split job market is what confronts immigrants and their ethnic children. They are further finding that a lot of opportunities are turning into contractual self-employment, many of which turn into ethnic niches, for example, Latino limo drivers in New York and Taiwanese computer-game designers in Los Angeles.

The economic base of cities is increasingly determined by their infrastructure, educational and research institutions, community services, and cultural life. The talent and creativity of a city's workforce is its resource base. Richard Florida may be overplaying the role of the creative class in economic growth, but the education, skill, and diversity of a city's population are undoubtedly strong determinants of economic prosperity.[81]

Cultural pluralism and its associated ethnic diversity are marks of cosmopolitanism that attract global capital and talent. They are the resources of the new urban economy. In this economic order, ethnic economies and enclaves have a significant role. They serve as platforms for global networking and become conduits for investment and labour from abroad. They contribute to the attractiveness of a city.

Chinese economies may be a special case at the present time because of their connections with the economic powerhouses of mainland China and Taiwan. It may not be the case for ethnic economies which are linked with economically weak countries such as Portugal, Mexico, or Senegal. It appears that ethnic economies will vary in their roles depending on such contingent factors and the social standing of the respective ethnic groups in particular cities. For example, the Asian ethnic economies yield more income than those of Blacks. San Francisco's Chinatown pays higher wages than New York's.[82]

Conclusion

Multiculturalism injects ethnic and racial alignments into urban economies. Ethno-racial diversity combined with civil rights and freedoms, the elements of multiculturalism, promotes ethnic entrepreneurship and advances the development of ethnic economies. Overlaying the class segmentation of economic opportunities, ethnic and racial groups come to dominate in certain occupations and economic activities. These clusters are the ethno-racial alignments in the economic organizations of multicultural cities. Such expressions of multiculturalism have been found in Toronto, New York, and Los Angeles.

The topics addressed in this chapter have included the organization, evolution, and role of ethnic economies, the basis of ethnic entrepreneurship, and differences in the earnings of various groups. These are the indicators of the embeddedness of ethnic and racial differences in urban economies. The observations on these scores are summed up above in the section entitled "Reprise: Ethnic Alignments in Urban Economies."

Ethno-racial diversity embedded in urban economies shows remarkable similarities in the three cities. The ethnic strategies of the respective groups are similar, the ethnic segmentation of consumer markets is common, the occupational concentrations of different groups have parallels, the evolution of ethnic entrepreneurship from dealings in ethnic goods to the trade and production of mainstream goods and services follows similar paths, and the boom in the development of ethnic malls and business centres is common. The suburbanization of ethnic economies has shown similar trends. The earnings of the non-White and foreign-born labour force were lower than those of the native-born and Whites for comparable human resources. The structure of inequality persists despite ethnic entrepreneurship. The differences among the three cities arise from variations in their opportunity structures and economic bases. Toronto's role as Canada's transportation hub and manufacturing centre has incubated ethnic niches. New York's health, educational, and financial services, combined with its tourist industry, provide opportunities for ethnic clustering. Los Angeles's role in the Pacific trade has used ethnic economies as the platform for overseas connections and investments.

Ethnic economies are fully immersed in local, regional, and national economic institutions. They exist within the confines of laws, rights, and policies. Their fate is inseparable from that of the regional economies and national business cycles.

Ethnic economies are symbiotically related to ethnic community organizations. They draw on social relations nurtured in language schools, religious institutions, and cultural organizations for business and in turn promote the development of community.[83] They are intertwined with ethnic media, cultural institutions, and political power.

The integration of ethnic economies in a regional economic organization is a challenge for policymakers. Ethnic economies need considerable leeway to realize their entrepreneurship and creativity. Yet they have to be prevented from fragmenting the local economy and must meet occupational safety, health, and fair wage practices. Again, reconstructing the common ground of fair competition, consumer protection, and workers' rights has to proceed in tandem with the promotion of ethnic economies.

To conclude, it can be said that ethnic economies open a path to employment for immigrants lacking in English (and French in Canada) language and other skills necessary for the mainstream economy. They help build up the assets of the self-employed and entrepreneurs and help in economic growth. They serve as the conduits for trade and networking abroad. They serve as a platform for innovation and entrepreneurship. Despite the current celebratory tone in the literature about ethnic economies, their being relatively inward-oriented is a matter of concern. They have to be more integrated in regional economies, without losing their inventiveness and entrepreneurship – a matter of delicate policy balancing.

Chapter Six

The Patterns of Community Life

The Terrain of Social Organization

On the surface, what differentiates a multicultural city is the variety of people encountered in its streets, shops, workplaces, and neighbourhoods. The echo of many languages spoken in public places, the smell of foods, and the strains of music from distant lands filtering out of homes and stores are the common hallmarks of a multicultural city. As one of the premises of our definition of multiculturalism suggests, all this variety is backed by civil rights that promote cultural and religious freedoms in social life. How is social life organized in a city where cultural diversity is recognized and ethno-racial equality is a right? How do ethno-racial differences affect its social organization? These questions are explored in this chapter.

Communities are the building blocks of social organization. They are formed by the interweaving of race, ethnicity, and class. Community formation is largely a process of the knitting together of people through social relations, which are often built around shared interests, identities, and values embodied in class, race, ethnicity, nationality, or religion. The clustering of people by ethnicity, race, or class, described in chapter 4, results in the formation of ethnic enclaves, racial ghettos, and gated citadels. They represent the ethno-racial contours of a city's social organization. The emergence of ethno-racially differentiated neighbourhoods as well as similar but non-territorial communities raise the issue of social segregation and a concern about the development of parallel societies within a city. This chapter will examine the patterns of segregation and assess their impact on the social organization of a city as a civic society. It will also reflect on the unifying role of the city as a civic society and culture across sub-cultural differences. Altogether, these questions will illuminate multiculturalism's impact on patterns of social relations, drawing on the experiences of the three cities.

The celebratory note in the description of multicultural cities should be qualified by the acknowledgment of undercurrents of racism and outbursts of ethnocentrism that lie below the surface of everyday civility. Racism in particular has a long history in North American cities. Though in the post–civil rights era overt forms of racism have been largely curbed in the public arena, racism still has some influence on the political and economic standings of ethno-racial groups.[1] A city of differences is particularly vulnerable to stereotypes and discrimination. Blacks, Natives, Latinos, and Asians, namely non-Whites, have been subjected to racism historically and sporadically even now. Skin colour and physiognomy mark an indelible identity that will not "melt" regardless of how long people have been natives of North America. Their inclusion and integration in the evolving national identity is the challenge of multiculturalism.

Social relations in a city are organized around a series of interrelated institutions, such as family, community, work, religion, and education in the form of persistent patterns of interactions. This is its social organization. Four sets of interrelated institutions are the primary channels for determining where and with whom people live, work, and relate to and how they gain social standing and status: that is, (1) housing and neighbourhood, (2) social network and community, (3) business and employment, (4) politics and law.[2] In this chapter, the first two sets of these interrelated institutions, namely, housing and neighbourhood, as well as social networks and communities are the focus of study. These two sets of institutions are the primary vessels for organizing social relations. The other two are more contextual in influence on social organization and they are extensively discussed in chapters 5 and 8.

The Common Characteristics of Urban Social Organization

Social relations in multicultural cities are not only organized around the shared ethno-racial identities and interests, but are also embedded in the economic and social order and contemporary modes of urban living. The latter affects the quality and forms of interactions in communities. To clarify the influence of urbanity, I will briefly recount the common elements of modern urban institutions.

Identifying urban ways of life, which include both social relations and cultural patterns, has been critical to the concerns of modern social theorists such as Emile Durkheim (1893), George Simmel (1902), Louis Wirth (1938), and, recently, Harvey Cox (1965), Manuel Castells (1983), David Harvey (1985), and Henri Lefebvre (2003), among others.[3] A discussion of their respective arguments and conclusions would not only be too long, but also tangential to the point that we are pursuing here, namely, how ethno-racial differences play out in multicultural cities. The following is a summary of the defining features of urban social organization as deduced from these theorists.

1. Social relations in a city are organized around institutions and activities. They are segmented, secondary (partial involvement), driven by interests more than personal ties, purposive and largely impersonal, and embedded in organizational norms and institutional mores, except for primary groups. Mike Savage, Alan Warde, and Kevin Ward, in reappraising Louis Wirth's theory of urbanism conclude that "although settlement type does not directly generate particular types of social relations, the frequency, density and context of personal contacts does have an effect on socialization."[4]
2. Heterogeneity of social backgrounds and roles is a defining condition of urban living. It produces individualism and interest-based groups. Ethno-racial differences add another layer in this edifice of heterogeneity.
3. Urban social structure is primarily based on class, in which social status is tied to economic standing and power more than to clan or family background. Class has a pervasive influence in determining how one lives, what life chances one has, with whom one associates, and what one's standing in society is. In modern times, class is defined by income, occupation, and consumption-lifestyle. Now there is a talk of knowledge and creativity as the determinants of class.[5] Social class permeates into ethnic and racial communities, defining status differences within such groups.
4. Collective goods are the wirings of a city. In their pure form, they are indivisible and in-appropriable, that is, they cannot be produced and consumed on an individual basis. Instead, they serve a whole community and thus have common behavioural norms and values. Manuel Castells, a Marxist sociologist, explains the urban problematic as the inequality of collective consumption, defined as "accessibility and use of certain collective services," among various

classes and their neighbourhoods.[6] Norms and values of collective consumption constrain ethnic behaviours and build a civic culture.

5. Finally, the social life in a city is divided between the public and private spheres. Public activities are contractual, involving little investment of emotion and memory, and are regulated by common laws, norms, and ethics with considerable situational improvisation.[7] Private activities largely take place with people one knows in a closed system of personal ties and reciprocal obligations. This sphere of social relations, with its norms and values, lays the ground for subcultures to flourish in families and communities, but it functions within the bounds set by the public sphere.

These are the structural conditions that underlie ethno-racial communities and groups. They colour social relations within ethnic and racial communities. Multiracial and multi-ethnic urban areas are affected by various processes of social differentiation, ranging from stratification, clustering, and concentration and to the structures of inequality, namely, segregation and polarization, ghettoization, and gentrification.[8] Cities differ in terms of the prevalence and scope of these processes.

Family, Neighbourhood, and Social Relations

Where one lives is an entry point into the social life of a place. From a home a network of social relations radiate out; from casual but regular contacts with store operators, daycare workers, teachers, and neighbours to intimate and persistent relations with friends, relatives, and co-workers.

These relations are mediated through class, racial, and ethnic identities, founded on shared values and interests, and affect where and with whom one lives and associates. Urban areas in the United States and Canada are initially organized as neighbourhoods and districts of the poor, middle, and upper classes through the workings of the housing market and public policies. Overlaid on top of these class contours are concentrations by race, ethnicity, and lifestyle. There are neighbourhoods where Blacks, Whites, Latinos, or Chinese dominate, as there are communities carved out as Jewish, Anglo, French, or Korean. Of course, a city has large areas of no particular ethnic/racial identity, where people of many backgrounds live side by side. This geographic sorting out of people by class, race, ethnicity, and lifestyle is a ubiquitous process of residential distribution in North American cities.

There are many forces driving the social sorting process: the housing market and income distribution, accessibility to jobs and services, social choices, and neighbourhood quality. And where one lives has a bearing on one's quality of life and network of social relations. As the phrase goes, "place matters."[9] Yet social and spatial segregation of the poor as well as racial and ethnic minorities has been a potent force in organizing community life in urban areas.

As for the structure of social relations in neighbourhoods, the family is the pivot around which they revolve. Differences in ethno-racial cultures as well as variations of the circumstance of immigration are reflected in various groups' family structures, which in turn affect the form and scope of social relations in neighbourhoods.

How do family structures vary from group to group? This is illustrated by the incidence of single-parent families in New York, which in 2000 ranged from high (32%–39%) for American Blacks, Puerto Ricans, Dominicans, and West Indians, followed by South Americans (28%), down to Chinese, Russians, and native Whites (12%–14%).[10] The Los Angeles area presents another variation in family structures. The county now has Hispanics as the largest group, 47.7% of the population. Hispanic families are often divided by split family migration, with one adult initially migrating and others following later. There are many truncated families made up of relatives and friends as boarders. In Toronto, ethno-racial differences in family structure mirror the US patterns. In 2001 Black racial groups, namely, of African and Caribbean origins, had high rates of single-parent families (31.4% and 28.4% respectively), while South Asians (Indians, Pakistanis) had the lowest rate of 5.5%, with East Asians (Chinese, Filipinos) slightly above them at 8.1%.[11]

Neighbourhoods with large percentages of single-parent families who are poor have fewer working adults, more unsupervised children, particularly young boys, and a relatively weak community organization and social capital.[12] These attributes are often the makings of ghettos, where there may be some social solidarity but a pervasive deprivation.

The single-parent family is only one type of family structure. Families can differ in other ways: number of children or seniors, single or multi-generational, joint or nuclear. Immigration introduces new family forms such as elders or adult children living with nuclear families. These differences have a bearing on the living conditions and need for schools, libraries, parks, community services, and other public goods, which give rise to neighbourhoods of divergent cultures, quality of life, and social relations.

New immigrants needing support and connections tend to converge in co-ethnic neighbourhoods, where they can get companionship, advice, and culturally appropriate goods and services. Middle- and upper-class immigrants may not necessarily need to be surrounded by co-ethnics or be so dependent on community services, though their non-territorial social networks often are with people of similar backgrounds.[13]

Yet these variations of neighbourhood structure occur within the parameters of what has been described above as the urban social organization. For example, in the cities of North America, conditioned by the pervasive values of individualism and privacy, neighbourliness is a relationship of politeness and structured informality. Neighbourliness means reciprocal relations with a selected few of similar lifestyle but a sharing of interests with other residents in schools, safety, and child care. The sociological theme of "loss of community" in cities is an expression of this limited scope of neighbourly ties and reduced emotional involvement in residential communities. Ethnic communities are also affected by this urban culture. A certain degree of formality seeps into their neighbourliness, though they show more interdependence and social solidarity than long-established areas of native-born northern Europeans.

Residential Segregation and Community Structure

The formation of ethnically and racially distinct neighbourhoods arouses concerns about the spatial and social segregation of minorities and immigrants. These concerns are particularly acute in view of the long history of discrimination and segregation of Blacks. Also, the goals of integrating and acculturating immigrants are assumed to be hindered by their living in ethnic neighbourhoods. Most of the late-nineteenth and the twentieth centuries' development of American and Canadian cities proceeded by housing immigrants from Europe and migrants from rural areas, particularly Blacks from the southern states in US cities, in the older districts of central cities. Thus, the Irish, Italians, Poles, and Jews were at various times viewed as "foreigners" and confined to the slummy parts of cities.[14] These experiences cast residential segregation as a social problem.

Over time European immigrants were assimilated, became part of the mainstream, and moved to more salubrious parts of cities, while non-Whites continued to be segregated in ghettos and ethnic slums.[15] Residential segregation was institutionalized through discriminatory zoning by-laws and city policies, housing-market practices of refusing to rent or sell houses to the discriminated groups, the redlining of ghetto areas by bankers, and housing covenants prohibiting selling to non-Whites or immigrants. Post-1964 civil rights legislation enforced fair housing practices and prohibited racial discrimination in the public sphere in both Canada and the United States. A new ethos of non-discrimination and open housing markets took hold in North America after 1965. The literature promoted the ideal of integrated neighbourhoods, though cities continue to develop areas of distinct class, ethnic, and racial identities in the suburbs and newer parts of metropolitan areas. The ideal of geographic integration has been articulated in the spatial assimilation model.

The Spatial Assimilation Model

This model's view is that as immigrants climb up the socio-economic ladder and are acculturated in the mainstream's ways, they transform their newly acquired economic and cultural resources into better housing and integrated neighbourhoods. Douglas Massey and Nancy Denton describe this as the process whereby "minorities attempt to convert their socioeconomic achievements into improved spatial position which usually implies assimilation with [the] majority group."[16] Ceri Peach has succinctly expressed the assumption of this model that "the social melting pot also melts the spatial enclave."[17]

This model is built on theories of assimilation that postulate the sociological washing away of ethnic and racial differences and project a modernist social organization that is primarily differentiated along class lines. Also, the mainstream society into which minorities assimilate is assumed to be culturally homogeneous and open to all. These assumptions do not bear scrutiny in the light of the historical experiences of minority communities and the stratified structure of the mainstream, as discussed in chapter 2.

The application of the spatial assimilation model to understanding the place of different racial and ethnic groups in cities has yielded mixed results. In the United States, middle-class Blacks do not seem to assimilate spatially to the extent the model suggests; though in Canada their relatively dispersed initial settlement lay the ground for further dispersal with the improvement in their socio-economic status.[18] Generally, the spatial assimilation model reflects the settlement experience of European immigrants and not of native Blacks,

Hispanics, or Asians. Similarly, it has greater traction in the United States than in Canada.

Alejandro Portes offers "segmented assimilation" as a refined model of the assimilation process. It postulates that assimilation occurs in specific strata or segments of the mainstream society that correspond to the social background of an immigrant.[19] Working-class Chinese, for example, settle in working-class inner-city Chinatowns. By contrast, Chinese professionals and business immigrants directly settle into the middle-class life of ethnoburbs such as Flushing in New York, Monterey Park in Los Angeles, Richmond in Vancouver, and Markham in Toronto. Even the second generation of immigrants assimilates upward, downward, or horizontally, depending on their social standing.

Furthermore, the choice of residential locations for immigrants and minorities is also influenced by preferences for the type of neighbours they want to live with as well as available housing and commercial, cultural, and religious services. The formation of ethnic neighbourhoods in contemporary cities is not just a matter of enforced residential segregation, but is also an expression of the preference for ethnic community life. Of course such choices are available more to the middle and upper segments of ethnic and racial groups.

One has only to drive through the booming suburbs of Markham, Brampton, and Mississauga in the Toronto area to witness the vibrancy of ethnic neighbourhoods. There are crowded ethnic malls, mega mosques and temples, circles and crescents lined with well-tended lawns and relatively new homes, and stores bearing multilingual signs. Social advancement has driven these ethnics/immigrants to stamp their identities on these areas and not "melt" in the mainstream. Nancy Foner's edited book *One Out of Three* profiles the social life of seven ethnic/immigrant groups in New York who have forged ethnic neighbourhoods in areas long known for their European heritage, such as Brighton Beach, Flushing, and Washington Heights. Los Angeles County has eighty-seven municipalities, and its neighbourhoods are little villages surrounded by highways. Immigrants, by moving into these places, have filled them with their businesses and cultural institutions and are beginning to run local councils. In the three cities, there is striking evidence of the formation of ethnic enclaves (also discussed in chapter 4), which are not the result of any systematic discrimination. On the surface, these areas represent the spatial segregation of ethnics/immigrants, but do they segregate them socially? Are these precursors of ghettos?

Residential Segregation: Ghettos and Enclaves

A ghetto evokes the image of a place segregated by race, racked by poverty, crime, with dilapidated homes, and broken facilities and services. Residential segregation alone does not make a ghetto, but its combination with economic, social, and physical blight earns this name.

The image of the ghetto is often raised to highlight the ills of residential segregation and its consequential social isolation. Such ills need not be the outcome of segregation, particularly if segregation helps build community institutions, preserve ethnic culture, facilitate religious observances, and incubate ethnic entrepreneurship. This is often the case with ethnic and immigrant enclaves. Frederick Boal argues that residential segregation can have bad consequences, but "in specific circumstances [it] may have positive functions."[20] What needs close attention are the conditions under which segregation takes place.

It appears that three conditions produce the prototypical ghetto: (1) involuntary exclusion and discrimination; (2) racial-ethnic segregation; and (3), within a racial/ethnic group, the isolation of the lower class. Neighbourhoods such as Harlem, the South Bronx, and Bedford Stuyvesant in New York, and Watts and East LA in Los Angeles have been the historical Black ghettos in the United States. They were the outcomes of discriminatory practices in housing and the job market in the twentieth century and a legacy of slavery, namely, a combined result of racial segregation and social exclusion. Furthermore, the flight of the Black middle class from these neighbourhoods turned them into areas of the underclass that lack leadership, role models, and connections to the city economy. Yet these places have always had a robust Black cultural life. Often residents found them to be struggling but supportive communities.

Recently, Harlem and Bedford Stuyvesant have been caught in the updraft of New York's real estate market. They are attracting affluent households of many races, which thereby is setting in motion a process of their transformation into fashionable places. A similar process of gentrification is under way in Watts, Los Angeles. With discrimination, poverty, class isolation, and blight decreasing, residential segregation alone ceases to make a ghetto.

Canada has not had US-style Black ghettos. Its race question has been largely concerned with the status of non-White immigrants (called visible minorities in Canada) and Aboriginals. Even its Black population is largely made up of immigrants from the West Indies and Africa, who are dispersed among other groups. Alan Walks and Larry Bourne, cited in chapter 4, observe about Canadian cities that "there is little evidence of ghetto formation along the US lines."[21] This observation is consistently confirmed by many other researchers.[22] While this does not mean that Canadian cities are free from residential concentrations of poor minorities, they seldom rise to the standing of being ghettos.

Undoubtedly, Toronto, Montreal, and Vancouver, the three premier metropolitan areas of Canada, show the convergence of two trends: (1) the growth of immigrant neighbourhoods; (2) a geographic concentration of poverty. The interweaving of these two factors results in the emergence of low-income immigrant neighbourhoods.[23] These ethnic neighbourhoods of poor immigrants are found in the rental apartment clusters that were built in the inner ring of the Toronto city. Yet they do not replicate the conditions of historic Harlem or Watts.

Apart from the various internal pull-and-push factors, some strong structural forces external to the system of urban community formation are leading to the concentration of low-income immigrants in ethnic neighbourhoods. The national and urban populations in Canada are growing primarily through immigration, which means that most of the new households are those of immigrants. The continual flow of immigrants is not only adding new ethno-racial households, but also swelling the ranks in the lower rungs of the economy, where large numbers of immigrants begin their life in Canada. Thus, there is a continual supply of low-income, new-immigrant households whose members choose to live in neighbourhoods with co-ethnics and where affordable housing is available. Does it mean that these neighbourhoods are turning into ghettos?

The answer for Canada is largely "no." There is a mix of middle- and lower-income households as well as a combination of renters and homeowners in ethnic neighbourhoods. Furthermore, these places have considerable internal diversity of national origins, languages, and cultures and, of course, incomes and education, even if one group dominates (see chapter 4 about ethnic enclaves in Toronto).

In recognition of these differences, the Canadian discourse on residential communities tends to use the term enclave and not ghetto. Ethnic enclaves are communities formed by choice that develop a thick web of institutions. Even poor immigrants have hopes of a good future. They feel they are on an upward trajectory, an expectation that is fulfilled for many over time and usually by the second generation. Also they are not isolated from the middle classes of their own kind. For example, immigrants' homeownership rates generally converge towards the national rate after sixteen years, though there are variations by ethnic groups and the business cycle at the time of arrival.[24]

Daniel Hiebert, in a study of ethnic enclaves of visible minorities (non-Whites) in Toronto, Montreal, and Vancouver, finds that the new residential order in these cities is a multilayered structure. It consists of both mixed minority (more than one ethno-racial group) as well as one-minority-group enclaves. They are seldom exclusive. He found that "enclaves are not mono-cultural landscapes, barring a few exceptions."[25] Even in the United States, a distinction is being made between ethnic enclaves and racial ghettos. The former are places of hope and choice, while the latter are the products of poverty and discrimination. One has to feel the energy and vibrancy of Jackson Heights, Flushing, or Washington Heights in New York to realize their differences from yesterday's ghettos, the South Bronx or Bedford Stuyvesant.

The point of the above discussion is that the spatial concentration of an ethnic/racial group does not necessarily lead to social segregation and cultural isolation.

Community Structures and Ethno-Racial Differences

Ethno-racial groups have comparable but not uniform residential patterns. First, native-born Blacks and dark-skinned immigrants, particularly of low income, are more segregated and limited in their choice of housing. Race seems to track differently in the housing market than ethnicity alone. Blacks have greater difficulties in accessing decent housing, employment, good schools, and thriving neighbourhoods than Whites and other non-White (immigrant) populations. Undoubtedly there has been a measurable change in the segregation and discrimination of Blacks in the United States.[26] Yet the historical structures continue to have some influence.[27] Second, among other non-White ethnics and immigrants, Asians fare better than Hispanics, and within these broad categories, Japanese, Indians, Iranians, Cubans, and Chinese are less segregated than other Asians and Hispanics. The differences in segregation of broad categories of ethno-racial groups are reflected in table 6.1.

The Dissimilarity Index (DI) is the most commonly used measure of residential segregation. It is a demographic measure of the evenness with which two groups are distributed in relation to each other across sub-areas (for example, census tracts) of a city or region. For example, if Asians form 10% of a city's population, then by this measure their

Table 6.1 Dissimilarity index (DI)

Toronto CMA (2006)	New York City (2010)	Los Angeles County (2010)	Los Angeles city (2010)
Black/Anglo-Canadian (0.50)	NH Black/NH White (0.76)	NH Black/NH White (0.67)	NH Black/NH White (0.69)
Jewish/Anglo-Canadian (0.66)	Hispanic/NH White (0.61)	Hispanic/NH White (0.63)	Hispanic/NH White (0.65)
Chinese/Anglo-Canadian (0.60)	Asian/NH White (0.52)	Asian/NH White (0.50)	Asian/NH White (0.45)
South Asians/Anglo-Canadian (0.59)			
Portuguese/ Anglo-Canadian (0.49)			

Sources: US data from A. Beveridge, D. Halle, E. Telle, et al., "Residential Diversity and Division," in *New York and Los Angeles*, ed. Halle and Beveridge, 314, table 11.1. Canadian data from M. Qadeer, S. Agrawal, and A. Lovell, "Evolution of Ethnic Enclaves in the Toronto Metropolitan Area," 324, table 3.

Note: NH = Non-Hispanic

percentage has to be 10% in each of the census tracts or any other sub-area of the city to have a value of 0.0, meaning no segregation. Segregation is measured in relation to a reference group. Any value above 0.0 means some degree of concentration of a group in relation to the paired group. The higher the value of DI, the greater is the concentration; if it reaches 1.0, then that group is concentrated in just one tract. The DI values shown in parentheses in table 6.1 indicate the proportion of a group that would have to move to "make the distribution of that group the same over all the geographic units."[28]

Table 6.1 reveals many differences in the residential patterns of selected groups in the three cities. (1) The ethno-racial composition of the Canadian and US populations and their social standings are different. Blacks have had a small presence in Canada and the native-born among them are all the more a small and dispersed minority. In Canada, residential patterns are based on distinctions of ethnicity and immigrant status. Therefore, DIs are measured by ethnicity against the distribution of White Anglo-Canadians. (2) Blacks in the Toronto CMA (Caribbean and African immigrants mostly) are less segregated than Asians. The most concentrated group was Jews (DI 0.66), which is largely the result of their choice of community cohesion and not the reflection of

Photo 6.1 Water conservation appeal for a multilingual community, Monterey Park, Los Angeles (courtesy Susan Qadeer)

any systematic discrimination. (3) New York is the most segregated city for Blacks (DI 0.76), followed by the city of Los Angeles. Suburban Los Angeles County is relatively less segregated (DI 0.67). Yet Blacks are the most segregated residentially in the two US cities and least segregated in Toronto (DI 0.50). (4) Hispanics in Los Angeles and New York are the second most segregated group. There is not a sizeable presence of Latinos in Toronto yet and they have no notable enclave. (5) Asians (Chinese and South Asians) in the Toronto area are residentially more segregated (DIs 0.60 and 0.59) than in New York and Los Angeles (DIs 0.50 and 0.45). They have large thriving enclaves in all three cities.

The secular trend is towards a decrease in the residential segregation of Blacks in New York and Los Angeles over time. There has been a slight increase in the segregation of Hispanics in both cities since 1980. Andrew Beveridge, David Halle, et al. have plotted DIs for these groups and conclude that "segregation remains relatively high in Los Angeles and New York regions, though it has declined in both, but more so in Los Angeles."[29]

Ethnic enclaves do not suffer from the historic drawbacks of segregation. Canadian residential segregation takes the form of enclaves, which are also appearing as the areas of immigrants in the United States. There is a considerable class, nationality, and religious diversity within enclaves in both the United States and Canada.

Canada has a better social safety network and social service system, including public health care, and thus its poor ethno-racial neighbourhoods are not places of despair. The housing conditions in Canadian cities are generally good. In 2006, only 2.1% of Toronto's housing stock was inadequate (requiring major repairs) and 5.1% was below suitability standards (not enough bedrooms for the household); but 12.5% did not meet the affordability standard of housing expenses, that is, more than 30% of the household income. The percentages are higher for rental than owner-occupied houses. Yet within these adequate housing conditions, poor non-White immigrants were mostly living in pockets of old apartment towers and townhouses in the inner suburbs, which have been described as deprived neighbourhoods.[30] Public housing in Toronto, unlike that in New York and Los Angeles, has a very diverse population. Its residents may be mostly non-White, but being immigrants they come from many different nationalities and ethnicities. For example, the Jane-Finch public housing complex, the largest in the Toronto area, has 138 nationalities speaking as many languages.

Asians do well generally in both Canada and the United States. They form enclaves and even when living in integrated neighbourhoods, they develop community networks, organize associations and institutions, and have strong family ties (as discussed earlier). Min Zhou et al. have charted the community life of Chinese in New York and Los Angeles. They describe the social development of Chinese communities in suburbs, documenting the development of Chinese American associations, academic tutoring centres (*buxiban*), college preparatory services, Buddhist temples, political organizations, and theatre and music academies in Flushing and Sunset Park, New York, and the ethnoburbs of Los Angeles.[31] Indians, Japanese, and Koreans have similar forms of community development. Pakistanis form community associations, literary clubs, poetry reciting groups, branches of homeland political parties, and marriage match-making services. The point is that they may appear to be residentially segregated, but that is not an indication of their social isolation or deprivation. Also worthy of notice is the trend towards forming associations and civil society organizations as a means of community organization, which indicates that ethnic communities are not organized as local groups of intimate personal relations, but are structured in the idiom of modern urban social institutions, that is, purposive and interest driven.

John Logan and Weiwei Zhang have analysed the socio-economic and residential patterns of Asians in the United States, focusing on the metropolitan areas. They have two findings of relevance to the study of community structures. One, Asians community organizations differ by nationality and ethnicity and it is not enough to lump them all together. Two, Chinese and Indians, though equal to non-Hispanic Whites in socio-economic status, are residentially as segregated as Hispanic, while Vietnamese are as segregated as Blacks.[32] They observe that although there are variations in Asian national origin groups (with Vietnamese living in the least affluent areas and Japanese, Koreans, and Indians in more affluent areas), they live in "better than equal neighborhoods compared to whites."[33] They describe Asians' situation as "separate but equal," suggesting that segregation does not always mean inequality.

The sum total of the above evidence is that the quality of the community life and social organization of ethno-racial groups can be good, despite their residential segregation.

Ethnic Networks and Integrated Communities

Although ethno-racial groups differ from each other by their degree of residential concentration, in most cases a majority lives in mixed neighbourhoods side by side with others. Chapter 4 discusses this phenomenon under the concept of global neighbourhoods. For example, in the Toronto area about 48% Chinese, 50% South Asians, 41% Jews, 29% Italians, and an almost negligible percentage of Black immigrants lived in enclaves.[34] Obviously the rest were dispersed among others.

Logan and Zhang have measured the exposure of Asian nationalities to their own group in their neighbourhoods for the total metropolitan population of the United States. They have found that nationally, the average Chinese neighbourhood was 14% Chinese, but had 29% Asians of other nationalities, the average Indian neighbourhood had 7.3% Indian, and 19% other Asians, while Koreans were living in areas that were 5.8% Korean and 22% Asian.[35]

What is beginning to be visible in American residential patterns is the effect of the majority-minority demography of US cities, which is

Photo 6.2 Multi-ethnic shopping plaza, Toronto

changing the ethnic and racial composition of neighbourhoods. The number of Whites is proportionately declining, and residential integration increasingly means living with other ethnic and racial groups. There are not enough White households to fill new neighbourhoods. The US Bureau of Census projects that by 2044, the United States will turn into a majority-minority nation. Non-Hispanic Whites will be less than 50%.[36]

Of course, low-income Blacks and Hispanics in New York and Los Angeles continue to be exceptions to the foregoing observation about multi-ethnic neighbourhoods. All in all, the majority-minority cities of the United States and Canada are becoming multi-ethnic and multicultural. Residential integration will have to be more and more defined by the multi-ethnicity of an area's population.

How does living with others in a neighbourhood affect social relations? Does it foster primary social relations among different groups? These questions have long engaged sociologists of neighbourhood studies. Social relations in modern urban societies are organized around interests, values, and lifestyles, and are mediated through associations and institutions. Social networks of friendship and kinship are based on primary relations, but around these primary groups are communities of varying degrees of secondary relations and purposive dealings. One's friends, relatives, and colleagues often do not live in the same neighbourhood. Strong communities exist without propinquity. (See chapter 1 for a discussion of urban community.)

Urban communities are networks of obligatory bonds, shared interests, and common values and purposes. Ethnicity acts as an interest articulated in the community's shared language, customs, music and literature, sense of humour, politics, and identity. Religious bonds further reinforce these social networks. Persons of diverse racial and ethnic backgrounds living in integrated neighbourhoods do live in close proximity to each other, but their social bonds are often with people of similar backgrounds and interests in different parts of cities.

A city neighbourhood is a community of weak social bonds organized around common interests in local schools, parks, municipal services, and the ethics of mutual help in emergencies. It is not a community of pervasive strong bonds, though individually one may be close to a few neighbours.[37] Ethnic neighbourhoods also evolve over time to be places whose social life is organized through secondary relations of institutions and association rather than by the bonds of intimate primary relations.

When ethnicity combines with religion, a community's social life is more tightly bound with social institutions. Churches, mosques, and temples become regular gathering places, as do annual and seasonal prayer services, meetings, and parades. Social relations are anchored in shared beliefs and rituals, with an ethos of trust and mutual obligation. Such is the case with the Orthodox Jewish enclave of Midwood, Brooklyn, where Avenue J, "might as well be the main street of a shtetl (where on Friday after sunset) the street becomes a ghost town."[38] Community life here is "embedded in submission to the Jewish law."[39] Though not as concentrated and big as in Brooklyn, Orthodox Jewish neighbourhoods of similar social organization are found in areas around Bathurst Street and Steeles Avenue in Toronto and in West Hollywood, Los Angeles.

Arab and Somali Muslims, Indian Sikhs and Hindus, Italian and Latino Catholics, and American Evangelicals also have some religio-culturally defined communities, where social life revolves around activities organized by mosques, churches, or gurudawaras. In addition to these social networks, kin, caste, or clan groups, social and professional associations, language and cultural organizations, youth clubs, and web communities are also elements of contemporary ethnic community life in North American cities. It is not uncommon in multicultural cities of extensive ethnic commercial sectors to have most of one's needs, such as those for groceries, physicians, restaurants, plumbers, or car repairs, met through contacts in one's ethnic network. One could live almost completely in one's ethnic circle. To quote a long-settled and well-integrated (university professor) Egyptian-Canadian describing his social life in Montreal: "I feel I work in an international company and live in Cairo. I am surrounded by Egyptians."[40] It may not be much different for native-born Anglo-Americans, Jewish-Canadians, or German-Americans. Obviously such networks are not based in neighbourhoods.

Of course, there are social networks that are devoid of racial or ethnic bonds. Friendships are formed across ethnic lines, and interracial marriages are increasing. Work places are the primary sites of mixing people. Social movements sweep up people of all backgrounds. Political or professional associations cut across ethno-racial lines. Sports, music, and art bind people of diverse backgrounds in shared interests. These social organizations are mostly in the public realm.

The conclusion suggested by the above evidence is that ethnicity and race interweave with interests, values, and tastes to organize social relations and community life in multicultural cities.

Civic Society and Social Integration in Cities

The foregoing analysis suggests that a multicultural city has a thick mesh of overlapping but distinct communities of ethno-racial identities and interests both territorially concentrated as well as dispersed. But it also has social networks formed around secular interests, values, professions, and purposes laid over ethno-racial communities. Claude Fisher has formulated the subcultural theory of urbanism, in which he maintains that larger places not only have distinctive subcultures, but also promote a greater intensity of distinctiveness, because of the critical mass and frequent encounters among people of the same subcultures.[41] On the surface, such cities appear to be divided into social segments, but they have an overarching unity both organizational and cultural. This is true about Toronto, New York, and Los Angeles. What pulls together the subcultures of cities? This questions needs to be addressed.

Cities usually are coherent civic societies of a distinct spirit, narrative, and reputation. Their local society and culture are part of the national society, but have some distinct characteristics. Toronto has long stood for public order and puritanical mores, but now it defines itself by its diversity and variety.[42] New York has always been a city of immigrants, where freedom and assertiveness reigns and the promise of opportunity beckons. Los Angeles has been identified with the virtues of a spread-out city, leisurely living, and Hollywood dreams, though now it regards itself as a global gateway.[43] Undoubtedly, these are stereotypes, but they represent the popularly perceived images of these cities, which also infect ethnic communities. Indians in New York or Jews in Toronto, for example, carry the marks of their cities' spirits in their outlooks. At some level, they can be differentiated from their compatriots in other cities.

As discussed in chapters 1 and 2, communities within cities are not autonomous social organizations and their cultures have limited purchase. They are lodged in a hierarchy of social formations, ranging from local communities and associations to city/regional and national societies, whose institutions provide comprehensive cultural frameworks and foster public values. Societal institutions such as the law, education, governments, the economy, protection and security, infrastructure, and environment are parts of the city culture. They lay down values, goals, and norms that cut across community subcultures and define the scope and limits of community behaviours. Cities as societies and their corresponding civic cultures serve as the common ground that contains and channels differences. A few examples will illustrate this point.

In the three cities, there are municipal and provincial/state public health laws that trump ethnic traditions about the handling and preparation of food. Ethnic restaurants in Toronto, for example, are periodically inspected by the city's public health inspectors and they have to meet the city's standards to get a "pass" to continue operating. New York and Los Angeles have similar controls. Parking restrictions may be suspended on the occasion of an ethnic parade or religious celebration in a particular neighbourhood, as an accommodation, but the rules are uniform. All have to conform to them at the risk of a fine. Civil and criminal laws set norms for a lot of individual and group behaviours. Diversity is not a basis for difference in their application.

Civic culture also defines the rules and expectations of public behaviour and those override ethnic subcultures. From sorting garbage into recyclables and wet refuse for the city's pickup, to the prohibition of polygamous marriages and the application of child-protection laws to parents, to norms of behaviour such as not resorting to fist fights in personal disputes, civic culture brings about coherence and unity among people of different subcultures. There are even rules about burials and funerals that override some religio-cultural practices. The point being underscored is that civic society and culture are the binding structures in cities of multiple subcultures and communities. They are the common ground without which multicultural communities cannot thrive and cities cannot function cohesively.

Another significant part of civic culture is the municipal government and the services it provides that make for the quality of life in a city. Cities' collective goods such as parks, streets, architecture, museums, theatres, art galleries, historic monuments, bohemian districts, public transportation, and social services bring people out of their networks and communities, providing opportunities for encounters with others. They cultivate a sense of belonging and promote unifying narratives about a city. For example, the New York subway system is so extensive and widely used that it is credited for the "liberal" outlook of New Yorkers, as almost everybody encounters people of different races and ethnicities daily. In the same vein, local politics, to the extent it is open, promotes people's involvement in city affairs, another unifying bond.

Ethnic and racial solidarity does not command unquestioned loyalty. For example, the interests of employees diverge from those of employers in ethnic economies; new immigrants compete with second-generation ethnics for jobs. Employees find themselves underpaid, overworked, and without benefits. They begin to smart under a double burden. They are blocked from mainstream employment by their non-marketable skills and poor English, on the one hand, and by their exploitative work conditions, on the other.[44] As they find that communal solidarity does not deliver fair working conditions, workers begin to join mainstream unions and resort to labour activism within ethnic economies in North American ways. Asian Immigrant Women Advocates (AIWA) has organized Chinese garment workers in San Francisco, New York, and Boston to demand fair wages and working conditions from their co-ethnic employers, after the mainstream labour unions had failed to enrol them.[45] In Los Angeles, a national labour union (SIEU) drive known as "justice for janitors" got largely Hispanic workers better terms of employment. An Indian American young woman has organized New York's mostly Bangladeshi, Pakistani, and Indian taxi drivers in a welfare association that agitates against exploitive co-ethnic taxi companies. The point is that formal institutions for labour rights begin to modify the patron-client networks of ethnic employers and employees. This is how civic culture cultivates common interests that override ethnic loyalties. It becomes an instrument of social integration in multicultural cities.

Cities need to be strengthened and made responsive to the evolving demands of citizenship for all. With the coming of majority-minority cities, the mainstream ethos will change and more deliberate planning for civic culture may be needed.

Cultural Change and Social Institutions

Neither ethnic cultures nor the mainstream culture are static. Old practices are abandoned or modified. New behaviours are incorporated in daily routines. New notions, values, technologies, ethics, and norms are adopted. Immigration itself is a transforming force for the host society, but it also turns immigrants' lives inside out. They have to learn a new language, adapt to an unaccustomed climate and landscape, and adopt unfamiliar behaviours, beliefs, and laws.

In the foregoing sections, we have come across examples of how ethnic communities evolve from webs of kinship, place of origin, or religion-based ties to interest-based social networks and goal-oriented associations. Correspondingly, mainstream institutions regularly incorporate norms and practices of ethnic communities, as persons of different subcultures come to populate the broader social structure. All these changes cumulatively affect the patterns of social relations and community structures.

The family is an institution that mirrors the continual cultural changes in social life. Ethno-racial communities are equally swept by the cultural winds of change in family relations. An illustrative case is that of Pakistani, Indian, or Arab Muslim families. The patriarchal structure of these families and their mores concerning women's segregation come into conflict with the economic and social demands of life in North America. A change in women's role is often necessitated by the need for income and by city living; many immigrant women who never worked side by side with men start going out to work in stores, workshops, and offices. Those who had never negotiated public transportation and city streets without male chaperons have to learn to go out alone and deal with strangers. Young Muslim women begin to espouse gender equality after exposure to such values in schools and colleges. The second generation in particular is active in Islamic institutions and ethnic associations, some battling patriarchal practices and finding Islamic justifications for their feminist ideas.[46] All in all, the structure of family changes and women's roles continue to be redefined. Of course, the path of institutional change is not linear. It curves back as reaction strikes back and tradition reasserts itself. Yet change is unmistakable. Even what appears to be a conservative practice, namely, wearing the hijab (women wrapping a scarf around their head but leaving their face visible) is an innovation most immigrant Muslim women have adopted in North America.

The change in Muslim families is germane to the discussion of institutional change, because in the post-9/11 discourse, Muslim culture is regarded as resistant to change. Evidence from other groups – Chinese, Italians, and Indians – further confirms that community cultures and social organizations are not locked into homeland traditions , but are growing and evolving systems. They continually interact with the mainstream's norms, values, and social movements, contributing to the evolution of the common ground of multiculturalism.

For example, Chinese families progressively become more gender-equal, allowing children, particularly the native-born, more freedom, though inculcating traditional values of hard work and discipline.[47] Issues of dual identity, the generation gap, intergenerational discord, domestic violence, and divorce are arising and awareness increasing in line with the national trends. Gay and lesbian rights are beginning to percolate through Chinese communities.[48] Rates of interracial marriages are reportedly quite high among Chinese.[49] Similar trends have been reported among Jews, West Indians, and others. The point is that community cultures in North America are being hybridized. They are borrowing and exchanging cultural traits with one another and the mainstream society. The dynamics of cultural change steadily restructures social institutions.

Conclusion

From the multicultural perspective, ethnicity, race, and culture play a critical role in aligning the social organization of Toronto, New York, and Los Angeles. They affect the formation of communities, the development of neighbourhoods, and the patterns of social relations in these and other multicultural cities. Who lives with whom, how social networks are formed, what forms spatial and social segregation take and what their effects are, what the role of a city is as a unifying society and civic culture: these questions point to the processes through which diversity affects the social organization of a city. By probing these questions, this chapter uncovers the ethno-racial contours of the cities' social organization. The following is an overview based on the findings.

North American cities have long been organized into communities and neighbourhoods around class, race, and ethnicity. Contemporary ethno-racial diversity builds on these structures. It extends the range of differences among communities and neighbourhoods, adding nationalities, culture, and immigrant status as the basis of community formation. It promotes new forms of spatial segregation, that is, enclaves and ethnic institutions that give new meanings and functions to the process of segregation and transform its social outcomes. Spatial segregation does not correspond to social segregation for ethnics and immigrants in the three cities in the same manner as historic ghettos did.

Territorial proximity in a neighbourhood is not an assured means of promoting primary social relations and close-knit neighbourhoods in modern cities. Social relations in cities are mediated through institutions, activities, and associations, except for primary groups of family and kin. The venues for social integration in cities are workplaces, schools, public arenas, institutions, local government, and collective facilities and services. Narratives and symbols, including the ideology of multiculturalism, also are means of fostering social integration.

Lest it be thought that multicultural cities are divided into ethnic quarters, almost a majority of ethnics in each of the three cities lived in mixed neighbourhoods. While living in mixed neighbourhoods, people form social networks, communities, and associations around their identities, interests, and needs. Affiliations based on shared identities, culture, values, language, and religion are a common basis for the formation of social networks that are geographically spread out. These are "communities without propinquity" in contemporary cities. Ethnicity and race as the foundational determinant of identities turn into interests and institutions around which social life is organized in multicultural cities. But these are not the only basis of social networks. Values, tastes, lifestyles, and occupations also form a thick mesh of overlapping social networks in multicultural cities, each serving some aspect of the multi-faceted identities of people.

Specifically, the chapter reveals the following. (1) There are parallel institutions and processes of community formation and ethnic/racial relations in the three cities. (2) Race matters. Native-born Blacks and dark-skinned immigrants are the most segregated groups, not only from Whites but also from Asians, in New York and Los Angeles. Canadian cities do not have that level of segregation by race; ethnicity is the basis of segregation. (3) Within broad categories of race and ethnicity, there are wide differences in family structures, degree of concentration, and community organization by nationality. For example, the Japanese are the least segregated residentially, followed by Koreans. Indians are as prosperous as Whites, but relatively more concentrated geographically in the metropolitan United States. (4) Contemporary urban social organization and norms of collective consumption also realign social relations in ethnic communities. In working-class ethnic communities, such as old Chinatowns, kinship ties, village/homeland bonds, and informal social organizations are the basis of social relations, whereas in middle-class enclaves and communities, formal associations, institutions, and interest-based social networks are the organizers of social relations. Even within ethnic communities, some degree of "relations by rules" and institutional norms comes to be practised.

The differences between the US cities of New York and Los Angeles, on the one hand, and the Canadian city of Toronto, on the other, lie in the social policies and ideologies of the two countries. They also arise from the differences in the local governments and civic cultures of the respective cities. Toronto does not have Black ghettos. Asians are more residentially segregated in Toronto than in New York and Los Angeles. Territorially, the most segregated group (DI 0.66) in the Toronto area are Jews, who are (relatively) prosperous, socially integrated, and White.

Spatial segregation does not mean social isolation for ethnics. Toronto's better housing stock, public health care, and services for immigrants are elements of its civic culture that contribute to social cohesion. Toronto has official multiculturalism, which is top-down. New York's and Los Angeles's diversity are sustained by private initiatives, but the stronger American national narrative and distinct civic cultures have a strong unifying influence.

Civil society and the civic culture of cities, be it Toronto, New York, or Los Angeles, act as unifiers. They cut across ethnic subcultures and norms, values, and regulations for public behaviour. At present the civic culture of the three cities is dominated by the historically rooted Anglo-European culture. There are also universal values that form part of this civic culture. With the emergence of majority-minority cities, the shrinking role of the dominant groups will necessitate the rebalancing of civic culture and society. The mainstream will change and its evolution will require deliberate cultural planning to preserve and promote universal values and ethics of shared citizenship. The construction and reconstruction of civic society and culture is as much a part of building cohesive and functioning multicultural cities as the recognition of identities and accommodation of diversity. This also upholds the two-sided model of multiculturalism developed in chapter 2.

Chapter Seven

Experiences of Living in Multicultural Cities

In this chapter, the aim is to describe the experiences of living in a multicultural city, where differences of culture and identity abound. These experiences are imprinted on the consciousness forged in the city and society. They do not form the totality of feelings and behaviours, but a subset that is a significant part of that consciousness. Specifically, the question addressed in this chapter is what are the distinguishing perceptions, attitudes, and practices imbibed from living with "others" in the same city? Also how are these perceptions and attitudes woven into the civic culture that cuts across differences? Simply put, how does it feel to live in a multicultural city?

The urban milieu has long been known to cultivate a distinct consciousness and way of life. As Georg Simmel wrote in 1903: "The metropolis exacts from man as a discriminating creature a different amount of consciousness than does rural life."[1] What he refers to as the "amount of consciousness" is a state of heightened awareness that comes from being in the presence of and in interaction with a large number of people who are markedly different. Multiculturalism adds another layer to otherness. Not only does it bring people of different race, culture, language, or religion together in a city, but also by honouring their right to be different it demands an acceptance and accommodation of others as co-citizens.

It has long been recognized in the sociological approach of symbolic interactionalism that "people use shared symbols to define and give meaning to their environment."[2] Normally this approach is used to observe the social meanings embedded in public spaces and activities. It can also be extended to images and meanings arising from exposure to social environments. Multiculturalism invests the social environment with the ethics of tolerating, accommodating, and accepting differences. One has to restrain one's prejudices, and cannot express or act on them without some social and even legal cost. Public life in multicultural cities is organized around the lived experience of accommodating cultural and racial differences. This point will become clearer as I discuss the experiences of living in Toronto, New York, Los Angeles, and other North American cities.

Exposures, Encounters, and Representations

In a city, exposure to different people, ideas, and modes of living is a defining experience. One is simultaneously being exposed to this rich diversity of stimuli and also exposing oneself to others. These exposures mostly take place in the public realm, but thanks to TV, radio, the Internet, and print media, they also filter into the privacy of homes. One strand of exposure, namely, to racial, cultural, religious, and linguistic differences, has a special bearing on the experience of living in multicultural cities.

In subways, on buses, on streets, or in parks, one comes across persons of different skin colours, accents, and customs, whose differences can be noted without affecting a necessary interaction. Richard Sennet aptly describes the reaction to exposure: "The eye sees differences to which it reacts by indifference."[3] Not that prejudice and racism are banished from multicultural cities, but they are socially delegitimized and legally suppressed. In a multicultural city, "strangers are us" and "the world in a city" are the metaphors that have a resonance. Exposure to differences is the first level of cultural-racial stimuli that condition the perceptions and reactions of the residents of multicultural cities.

A city ties together residents in webs of mutual interdependence and interaction. Through these inter-linkages, experiences of ethno-racial difference are woven into residents' everyday dealings. The experiences follow a path that leads from exposures to encounters, tolerance, engagement, acceptance, and relationships. This is an ongoing project that moves in a rhythm of two steps forward and one backward. Through all this, the ethics of mutual accommodation and cultural exchanges evolves.

Some mundane activities can bring people into contact with persons of different skin colour, accents, languages, as well as habits. A White man goes into a subway and finds a seat next to an African or a hijab-wearing woman, for example. This encounter begins with stereotypical images of each other, but the norms of being co-passengers demand mutual courtesy, which in turn humanizes their respective images to some extent. Such casual encounters occur on streets, in malls, on buses, and in other public places. These are sites of controlled encounters, and out of millions of such active and passive interactions emerge perceptions, attitudes, and practices of familiarity, tolerance, and accommodation but also, sometimes, hostility.

More sustained interactions in offices, schools, hospitals, places of worship, and clubs reduce social distance and build mutual trust. This is how differences are bridged, not washed out, and a composite civic culture is forged. The experience of living in multicultural cities is socially transforming, seldom dull. Everyday life in multicultural cities is a theatre of surprises and adjustments. This process is mediated by the laws, language, regulations, norms, and values of mainstream culture, which acts as the medium for interrelations.

Everyday Multiculturalism

At the level of everyday life in a city, multiculturalism comes into play in the ways residents communicate and interact across their cultural and racial differences. It lays the bases of an ethics of citizenship that binds diverse people together in civic reciprocity. In other words, everyday multiculturalism is about how cultural diversity is experienced and negotiated in day-to-day dealings.[4]

Everyday multiculturalism is enacted through behaviour in public places. The norms and values that regulate such behaviour are gradually stretched and modified as people bring to them different accents, languages, habits, values, and images. An Armenian grocery store in New York had a large handwritten sign that said, "I am not angry, I speak loud." Obviously, the owner had found out that his "normal" speech was annoying his customers. This little deviance from the norm for decibels is illustrative of not only how cultural differences affect everyday communications, but also how they are accommodated through mutual awareness.

The processes of everyday multiculturalism are observable in many social situations and institutions, such as in the narratives (including stereotypes) of different groups about themselves and others, in encounters and dealings, in consumption (particularly of food, music, and fashion), and in the diffusion of new norms in economic and political activities. Underlying these interactions and exchanges is the inter-ethnic politics of power and influence, which Ash Amin describes as the "micropolitics of everyday contact and encounter."[5] In the following sections, I will discuss various aspects of everyday multiculturalism unfolding in these situations and activities.

Narratives and Imageries

Narratives are the glasses through which people perceive one another as well as regard each other's histories and identities. Narratives about other groups are ideas and stories constructed out of images, ideologies, and the experiences of dealing with others. Stereotypes are simplified, even caricatured, images of others drawn from narratives. They condition perceptions and attitudes initially, but sustained inter-cultural encounters help to modify or at least reduce their hold. "Super organized Germans," "brainy Jews," "spiritual Indians," for example, are stereotypes that frame the initial perceptions about persons of such backgrounds. Yet when one meets them as neighbours, sales persons, or co-workers, one's perceptions about them are reinforced, humanized, or modified, depending upon one's experiences, both as an individual and in a group.

A multicultural city brings people face to face with persons of diverse backgrounds every day. These encounters cannot but affect perceptions and images, particularly as they occur under the umbrella of norms and laws of non-discrimination and reciprocal acceptance. And that is why the overarching civic and societal cultures are necessary mediums. Encounters and dealings may not necessarily lead to mutual understanding, but they require extending tolerance to each other. They also generate competition, a sense of superiority, and perceptions of deprivation or feelings of hostility, depending on the situation.

Responses to exposures and interactions with people of diverse backgrounds take two different forms, one at the individual and the second at the group level. At the individual level, encounters and dealings with others generally lead to mutual acceptance and accommodation, fulfilling the promise of the adage that interaction reduces prejudice. One learns to put up with strong accents, linguistic differences, and alien customs, thereby finding ways to carry out one's business with civility, because one must. By and large, at the personal level, a multicultural city produces civility and accommodation.

Out of thousands of daily contacts between persons of different races and cultures, only a few may go sour, resulting in angry outbursts.

Most interactions are purposive and their shared goals help bridge cultural chasms. In the process, a common set of symbols and meanings emerges that affect the behaviour of all those involved in interactions, for instance, the ethics of "not creating a public scene." Social contacts in cities are generally functional, impersonal, segmentary, and ephemeral. Such relations promote tolerance, because they do not normally carry any expectation of closeness. Casual contact facilitates mutual accommodation. Cumulatively, these interactions have a transformative effect on individuals' attitudes and behaviours. One learns to accept people with their differences and appreciate cosmopolitanism.

At the group level, cultural and racial differences are mediated by economic interests, political power, and ideological narratives. Encounters and dealings at the group level take many forms, from indifference, to alliances, political-economic exchanges, competition, hostility, and occasional outbursts of violence. A group is an abstraction. Its facelessness makes it a stereotypical category, based on ethnic, class, racial, or cultural characterizations. Images and stereotypes, positive or negative, play a stronger role at the group level than in dealings among individuals.

The history of White–Black relations in the United States is a vivid example of how evolving views about social rights have structured their dealings over time. A long process of changing narratives and evolving race relations marked the period from the American Civil War to the civil-rights laws of the 1960s. Canada has had its own struggle of redefining race relations from the colonial oppression of non-Whites and Aboriginals to present-day multiculturalism. Some remnants of these histories still exist in the form of racist practices in the United States and Canada, for example, racial profiling by police and employment inequities. Inter-ethnic relations have now evolved beyond Black–White dealings. There are distinct narratives of various ethno-racial combinations, such as Latino-Chinese, Korean-Blacks, South Asian–Chinese, and West Indian–African Americans. In a multicultural city, inter-ethnic relations at the group level follow many parallel tracks.

Inter-ethnic Relations

The tolerance and accommodation exhibited in inter-ethnic dealings at the individual level do not invariably translate into harmony at the group level. In New York, Los Angeles, and other US cities, Blacks and Jews had daily dealings as customers-merchants, welfare recipients–social workers, or tenants–landlords up to the 1970s. Koreans in particular and Asian immigrants in general have been replacing Jewish merchants and leapfrogging into professions and businesses, bypassing Blacks and Latinos. Though at an individual level they generally accommodate one another and maintain civility in relations, as groups their relations have an undercurrent of tension and suspicion. Jennifer Lee, in an ethnographic study of the relations between Jewish and Korean merchants and local customers in poor Black neighbourhoods of New York and Philadelphia, concludes: "While day-to-day interactions between merchants and customers may be civil and ordinary, small events can trigger anger with race polarizing the simplest interactions."[6] These tensions periodically burst out as merchants' boycotts, inter-ethnic violence, or riots. This happened in Los Angeles in 1992 at the time of Rodney King's killing by police and in the New York looting of Jewish stores in Harlem in 1968, the firebombing of a department store in Harlem in 1995, and the boycott of Korean merchants in Brooklyn in 1990.[7] Still, a few days of hostilities punctuate 350 days of civility.

The power of narratives and images in inter-ethnic relations at the group level can be observed in Korean employers' relations with Latino workers. Chong-Suk Han, in an ethnographic study in Los Angeles, found that Korean employers believed that Latinos were fit for muscle work only, and not able to calculate and make change.[8] This stereotype created a split labour market and caste-like structure of employment in Korean establishments.[9]

Negative stereotypes and unfavourable images are not the only narratives defining inter-ethnic relations. Positive images are an equally significant influence on relations among ethnic groups. Furthermore, narratives evolve over time, switching from negative to positive images (and vice versa), as ethno-racial groups collaborate with each other politically and develop complementary economic niches. Also, the image of a group changes as its status and prestige improves, sometimes by a change in the international standing of its home country. As China has become an economic powerhouse that finances the debt of the United States and drives global trade, the status of Chinese in the States and Canada has risen correspondingly. They are courted for investments and partnerships, particularly in real estate.[10] With the economic growth of India and the success of its software engineering industry, a similar change is under way in the prestige of Asian Indian immigrants.

The images of the Japanese and Chinese in both the United States and Canada have undergone major changes in the past decade. They are now regarded as a "model minority" and are viewed as economically entrepreneurial, academically brilliant, and socially adaptable. Their socio-economic profile bears out this image.[11] Because of their high grades in school, Asian students have come to dominate the engineering, science, and mathematics faculties of prestigious universities in the United States and Canada. *Maclean's*, Canada's leading weekly magazine, reported that the high admission rates of Asians in universities are causing concerns about an ethno-cultural skewing of campus life.[12] The scholastic reputation of Asians is helping to change their image.

White–Asian interactions are increasingly marked by understanding and civility, though incidents of racism occur. The relations between Asians and Blacks, Latinos and other ethnic minorities, are the new frontier of narratives. In majority-minority cities, new challenges of inter-ethnic relations lie in the interactions among immigrant Latinos, Asians, and others. White–Black relations are becoming one among many points of potential ethno-racial competition.

Immigrants are the primary bearers of cultural diversity in multicultural cities, though their second generation is also sizeable.[13] National attitudes towards immigrants get woven into the images of ethnic groups and consequently they affect inter-ethnic relations. A wave of anti-immigration sentiment in recessionary times brings out stereotypes about immigrants taking away jobs and not integrating into society. Conversely, in economically good times, inter-ethnic narratives turn positive. Similarly, international relations with immigrants' home countries have a bearing on attitudes towards them. The Iraq and Afghanistan wars and the terrorist acts of Muslim extremists gave rise to anti-Muslim sentiments. Japan's dominance of the US auto market in the 1990s and, in recent times, China's favourable balance of trade have sparked anti-Japanese and Chinese sentiments in high-unemployment regions of the United States and Canada. Such feelings raise suspicions and affect public discourse about particular ethnic groups.

The Consumption of Variety

Food, music, media, and art are the vehicles of everyday multiculturalism. Through them, people experience cultural diversity and communicate differences. They enrich everyday life by offering opportunities for experiencing a wide variety of tastes and sensations. The consumption of variety becomes a pursuit by itself and a source of excitement. It lays the basis for cosmopolitanism and cultural exchanges.[14] This form of everyday multiculturalism widens the range of common norms and behaviours. It contributes to the process of restructuring the common ground.

A multicultural city is distinguished by the choice of cuisines it offers. In the United States and Canada, successive waves of immigrants have brought their homeland foods, many of which have found a place in North American cuisine, for example, the German hamburger as well as frankfurter, Italian pasta, Mexican beans, and Chinese stir fry. What is striking about contemporary multicultural cities is that they stake their reputation on the variety and quality of their eateries. They assign a place of pride to the offerings of ethnic foods. San Francisco promises the best Chinese food outside China. New York is the home of cuisines from almost the whole world. Toronto promises a feast of ethnic tastes.

The popularity of eating out has brought out ethnic cuisines from homes to restaurants and into the mainstream. The food courts in shopping malls and office towers are living symbols of how "normal" ethnic food outlets are. The demand for a variety of foods is high enough to induce New York, Los Angeles, and even Portland to allow the selling of ethnic fare from food carts.[15] Toronto has been opening up slowly to ethnic food carts by tweaking its regulations.[16] The fare on offer usually includes Italian, Chinese, Greek, Mexican, Middle Eastern, and Indian dishes apart from the normal hamburgers and fries, sausages, and sandwiches.

The diffusion of foods produces a fusion of cuisines and convergence of tastes. In New York, Los Angeles, Toronto, and other multicultural cities, the tolerance of spices in food has increased all across the board. Almost as many Anglo-Americans/Canadians as the respective ethnics crowd into Indian, Hunan Chinese, or Mexican restaurants to enjoy "hot" dishes.

Of course, ethnic food is not always a promoter of harmony. Sometimes neighbours' complain about "stinking" spicy odours as ethnics move into apartment buildings or into houses built closely nearby. Still, the acceptance of new smells and tastes comes with exposure, even if the process is not always smooth.

At the same time, some cultural theorists and literary purveyors of authenticity look down upon the multiculturalism of food and festivals. Stanley Fish calls the excitement of ethnic restaurants and weekend festivals a "boutique multiculturalism" that is characterized by a "superficial and cosmetic commitment to diversity."[17] No one would claim that racism or ethnocentrism can be eviscerated as a result of enjoying food and festivals from many cultures. Yet encounters with the pleasurable aspects of others' cultures cannot but promote some appreciation of diversity.

Art and Media Multiculturalism

Music is another form of sensual consumption whose variety and diversity is a mark of multicultural cities. The musical repertoire of North America has always been open to the music of distant lands and ethnic communities. It includes many musical genres of foreign origin. African-American blues, Italian opera, and the Beatles' pop, for example, have thrilled generations of Americans and Canadians. Late-twentieth-century immigration and globalization have brought music from Latin America, the Caribbean, China, India, the Middle East, and Southern Europe of both the folk and classical traditions. Musically, North America has become a global village.

Our main interest here lies in the social consequences of the flourishing of ethnic and global music. A variety of music enriches the cultural and artistic life of a city. The fusion of different styles links together different groups and extends the common ground. A few examples of ethno-music that is being integrated into the North American repertoire include Mexican ranchera, Latin meringue, Bollywood Bhangra, Pakistani qawalli, African drum ensemble music, and French-Algerian raï. They have spilled out of their respective ethnic niches and have fused into mainstream music. Correspondingly, the mainstream's rhythms, harmonies, and instruments are being incorporated into the ethno-music.

The variety of music is closely tied to the development of ethnic radio and television outlets, which are part of the explosive growth of multicultural literary, artistic, and electronic expressions. To give an idea of how extensive are these activities, while not attempting to document their full scope, I will briefly recount the growth of some immigrant communities' media and literature.

There is a long tradition of ethnic newspapers and literary/artistic clubs emerging with the growth of immigrant communities. In the United States, Germans became the largest non-British group between 1875 and 1915. There were about 800 regular German publications, including newspapers, in the States by the end of the nineteenth century.[18] The *New*

Yorker Staats Zeitung was a daily German newspaper in New York. There were German schools, literary clubs, and singing societies in most cities. Jews founded Yiddish newspapers and publishing houses, a tradition that survives still in New York. Isaac Bashevis Singer, a New Yorker, won the Nobel literature prize for his Yiddish writings in 1978. So vibrant have been ethnic literary and artistic activities in New York that even before multiculturalism they had materialized as an ideology.

Multiculturalism has lifted ethnic music, media, literature, and art out of the niche of immigrants' cultures into the mainstream. Ethnic cultural expressions have a new legitimacy. Their spread has been aided by the new media of digital and satellite television, YouTube, videos, and Facebook. New York has 270 newspapers and magazines in 40 languages, over and above long-standing Black and Spanish presses. Just to take the example of Chinese media, New York has seven and Los Angeles five Chinese newspapers. Toronto has three Chinese dailies and many weekly community newsletters. There were about 200 Chinese media outlets of all types in the United States in 2006.[19] There are four Chinese radio and television stations in Los Angeles and New York respectively and one in Toronto. Apart from these local channels, these cities' cable companies offer bundled telecasts of Chinese television programs originating from Taiwan, Hong Kong and the mainland.[20] Anybody surfing television comes across Chinese news, music, and drama programs that stretch over twenty-four hours.

Similar is the situation for South Asians, Hispanics, Korean, and Italians. Toronto has two local multicultural stations offering programs and community news in twenty-two European, Asian, and Latin American languages. Its major cable network offers 190 TV channels in over 34 languages.[21] The Asian Television Network (ATN) offers thirty-three local and satellite channels for South Asians in various languages of India, Pakistan, Bangladesh, and Sri Lanka.[22] Similarly, the Commonwealth Television Network (CTN) has offerings from the West Indies, including a channel devoted entirely to cricket. In ethnic restaurants and bars, satellite TV programs from home countries are on display. One can live in New York, Toronto, or any other city and watch Brazilian, Greek, or almost any other country's television all the time. In these cities, it is now an everyday experience to walk into a bar, restaurant, or grocery store and find a soap opera or sport in a foreign language playing prominently on the TV screen. It is almost the new normal, far from the days when such a public display of another language was a sign of disloyalty. The point is that the ethnic media are now global in reach, and have become an integral part of the cultural institutions of North American cities.

Ethnic media, literature, and art enrich the life of ethnic communities, affirm their identities, and provide an outlet for their creative energies. There are poetry reciting meetings, literary clubs, painting exhibitions, and dance/drama performances organized in different languages. These are the expressions of ethnic cultures that often go unnoticed in discussions of multiculturalism. Not infrequently, they find a place in the mainstream's concert halls and theatres.

For society at large, ethnic cultural and artistic productions are a source of new ideas, sensibilities, and aesthetics. The variety of such offerings is by itself stimulating and exciting. In the creative cities of today that marry art with technology, the intellectual and aesthetic tensions that arise from the coexistence of different modes of thought are recognized to be incubators of creativity.[23]

The fusion of ethnic-global themes and ideas in North American art and literature has produced new literary genres in English. Latino authors James Diego Vigil, Susana Chavez-Silverman, and many others have brought new sensibilities to American literature, as has Richard Rodriguez. Chinese writer Maxine Hong Kingston won the National Book Critics award in 1976. Amy Tan's novel *The Joy Luck Club* was on the *New York Times*'s best-seller list for many weeks, has been made into a movie, and was translated into thirty-five languages. Indians writing in English have spawned a new literary form called Indian English literature. V.S. Naipaul, Salman Rushdie, Jumpa Lahiri, Kiran Desai, and many others have turned sub-continental themes into universal messages. The literature in turn inspires movies: *Slum Dog Millionaire, Monsoon Wedding*, and *Namesake* are recent North American box office hits universalizing Indian concerns. This is how multiculturalism is being woven into the mainstream. Philip Kasinitz et al. observe that "the creative mixing of immigrant and native minority cultures [is] already clearly evident in the music, art, dance and poetry being produced in hyper diverse cities like New York and Los Angeles."[24]

The above-discussed artistic and literary diversity and its creative impulses are largely expressions of middle-class multiculturalism. Similarly, the rhetoric of multiculturalism is largely the province of the middle and upper strata; the working class and the poor experience cultural and racial diversity in concrete personal ways. Anchored in their daily routines, their experiences are contingent on who they work with and live around. Power and status play a bigger part in their encounters with persons of different backgrounds. They are more deeply rooted in their kin, ethnic, and religious networks than the middle and upper classes. Their views, which tend to be stereotypical, are mostly confined to musings within their own circles, though their dealings with others are framed by their contingent interests. Overall, the multiculturalism of the working classes is negotiated on a personal and situational basis. Soccer or football fans are a good metaphor for the lower class's multiculturalism. They admire their favourite teams' players regardless of race and ethnicity, but are not restrained in calling out slurs if swept by a situational sentiment. Working-class multiculturalism is based on the politics of daily interactions.

Reconstructing the Social Order in Public Places

The reconstruction of social order in multicultural cities is a two-way process. On one side is the redefining of expected behaviours and practices to accommodate diversity; on the other is the affirmation of public norms and expectations for people of divergent backgrounds, particularly new immigrants. Public places are the venues for the reconstruction of social order; that is where people of divergent cultures and identities encounter each other and develop mutually binding expectations.

Recently arrived immigrants can readily be spotted on subways and buses. They are the ones who often crowd the doors getting in, without waiting for those coming out, and vice versa. The etiquette of boarding and leaving public transportation is soon learned after a few disapproving stares and by observing others. In the case of new subway lines, the norms of riding public transit have to be explicitly laid out. When Vancouver's new Canada Line was opened, the authorities printed instructions for the use of passengers. The city of Delhi, India, had volunteers patrolling a new subway line instructing passengers not to squat or spit on the floor.[25] Making one aware of public expectations is a step in constructing social order.

Jane Jacobs said that the social order of the street is "kept primarily by an intricate, almost unconscious, network of voluntary controls and standards among the people themselves."[26] This social order evolves in public places, such as streets, parks, plazas, theatres, and stores, through the (micro) politics of doubt and trust expressed among strangers.[27] In these politics, stereotypes and suspicions lurk below the surface, but shared experiences and frequent contacts, combined with the ethos of citizenship, resolve such feelings over time. The city's public realm induces tolerance and civility in everyday life.[28]

As a public activity, sports bring people together as players as well as spectators. Ethnic sports serve two functions. One, they build strong communal feelings for ethnic communities. Second, they draw together persons of different backgrounds in shared pleasures. Soccer is a case in point. Most national soccer teams in Europe and North America are made up of players of immigrant origins, namely, Latinos, North Africans, East Europeans, and Africans. Many national heroes are foreign-born. The World Cup of soccer is an occasion for celebrations in cities like Toronto, New York, and Los Angeles. Noisy parades and street parties are held at the times when ethno-national teams are playing a continent away. Yet these spectacles are open to everybody in the spirit of partying together, a spirit that occasionally also sparks brawls.

Cricket, popularized by South Asian, British, and West Indian immigrants in North America is becoming a public sport in the Toronto area, with 172 teams registered with the District Cricket Association.[29] New York City's department of education has added cricket as the newest league game, with fourteen teams. City parks on the edges of the Bronx (Van Courtland), Brooklyn (Canarsie Beach), and Queens (Baisley Park) are usually crowded with cricket players on summer weekends.[30] In the same vein, the Chinese Canadian Table Tennis Association of Toronto is a booming club, a quarter of whose members are non-Chinese.[31] Yoga is a practice that has been mainstreamed in North America. These examples show how activities originating from ethnic niches spread out into the public realm.

Stores, bars, and clubs, though private places, are functionally part of the public realm. In these places, wide latitude is given to one another's manners, accents, languages, and modes of living. Laws, business ethics, and self-interest bring about a convergence of cultural mores towards common norms. In North America the corporate sector leads in laying norms for business practices. McDonalds, Tim Hortons, Target, and Wal-Mart lay down models of customer service. Their staff may have varying accents and manners, but they cannot stray too far from company policies. New immigrants working in such establishments have to learn the North American ways of dealing with customers and co-workers. Family stores and ethnic businesses may have different mores, but they are continually pulled towards the practices of the mainstream. Ethnic restaurants have to pass public health inspections. Ethnic customers expect cleanliness and satisfaction. These are the ways in which social order is constructed and reconstructed. "A very Halal Christmas dinner" was the announcement posted on a school bulletin board in a neighbourhood with a sizeable Muslim population in Toronto.[32] It captures the spirit of an evolving social order.

Distrust of the Other: Competition and Controversies

Harmonious and responsive social order is always a work in progress in multicultural cities. Woven into the experiences of living in multicultural cities are undercurrents of racial and ethnic tensions, incidents of disorder, and the flaring up of cultural-religious controversies. Most residents may not personally experience any of these, but they are aware of them through the media, rumours, and stereotypes. Their positive feelings are tinged with lingering distrust of the other. Mark Hutter calls this the ecology of fear, expressed as "the concern for social order in a world of strangers."[33] The ecology of fear is given particular shape by the social institutions of a society, such as its ideology of race and ethnic relations, economic organization, and demography.

A social order could be well organized and stable, but divisive and oppressive. This was the case before the enactment of the Civil Rights Act (1964) in the United States and the Canadian Bill of Rights (1960) followed by the Charter of Rights and Freedoms (1982) in Canada. The present-day social order of interethnic relations is structured by the current laws and norms of "freedom from discrimination," the institutions of a post-industrial economy, and the imperatives of dependence on immigration for the labour force and population growth, as well as processes of globalization. These macro-forces affect the micro-relations of everyday multiculturalism. One may not overtly express discriminatory behaviour, but one could harbour fears of others sparked by economic and social competition. In pursuit of jobs, housing, safety, and space, people are prone to distrust. They look on competitors of different backgrounds through the imagery created by the media, at least initially. This is particularly true at the group level. As discussed above, individual interactions are marked generally by civility required by the social order, but at the group level distrust of strangers and negative imagery of others have resonance. These feelings spark community controversies and fuel the politics of inter-ethnic relations.

Racial and ethnic identities are signalled by skin colour and facial features, which in social narratives become the markers of images and stereotypes. These markers serve as cues for relations with strangers.

Racism is an ideology that prescribes attitudes of superiority and exclusion. In North America, such attitudes are mostly directed towards people of colour, Black in particular. Although overt acts of racism against individuals have been greatly reduced since the 1960s, practices of inequality linger in impersonal and anonymous dealings. These feelings have a bearing on inter-racial and inter-ethnic contacts in public places.

The association of crime with certain minorities has produced behaviours such as resorting to avoidance manoeuvres on coming across a bunch of non-White youth in the street. The fear of disorder is often palpable in situations of anonymity, though on a one-to-one basis the encounters are comfortable. At the institutional level, police profiling of Blacks is a common practice in almost all cities of North America. In New York, for example, the police department made 600,000 street stops in 2010; 84% were of Blacks and Latinos.[34] Toronto, Los Angeles, and other cities are similarly accused by minorities of singling out non-Whites in routine checkups, whether by police or immigration agents In post-9/11 times, Muslims have found themselves the object of police and security agencies' profiling. From the perspective of our discussion, the point of these examples is that the social order of everyday relations is not independent of the social structure in which it is vested.

In multicultural cities, community controversies are often flaring up among ethno-racial groups competing for political and social power, seeking expression of their religious and cultural symbols and practices in the public realm, and vying for equal access to jobs, housing, and services. To take the example of Korean merchants' conflicts with their Black customers and employees, discussed above, they had simmered for a long time. There were confrontations and boycotts in New York and Los Angeles, but the mutual distrust came to a boil in Los Angeles, resulting in the killing of a fifteen-year-old Black teenager and the shooting of a nine-year-old Korean girl in 1991.[35] These incidents were fuelled by merchants' perceptions that their Black customers were hostile and untrustworthy, while Blacks saw merchants as disrespectful, rude, and predatory.[36] Memories of such episodes freeze into stereotypes. The media plays them up, straining inter-group relations for a long time, though courtesy prevails in daily dealings between individuals in the contending groups.[37]

The process of change in the demography of a neighbourhood is often the wedge that divides established residents from newcomers. It ignites fires of inter-group confrontations along ethno-racial and/or class lines. An illustrative example is the Chinese-Italian contest for the control of Little Italy in Manhattan, New York. The neighbourhood's old housing stock had long been in decline. It was filtering down to the Chinese immigrants while Italians were moving to the suburbs. Yet the commercial core of Little Italy thrived as an Italian community's hub and weekend destination. In 1990, only 10% of the district's population was of Italian ancestry and another 23% were of mixed heritage, while a majority in the district was non-Italian, with Chinese in dominance.[38] The infiltration of the Chinese precipitated a long confrontation about the control of the neighbourhood. The New York City Planning Commission designated it a special district to preserve its cultural character. The Chinese called it a plan to push them north of Canal Street.[39] Today Little Italy is limited to a few blocks of Mulberry Street and survives as a commercial and cultural place in the midst of Chinatown. The intercommunal disharmony has diffused as Chinatown itself has evolved into the market for knockoffs of brand-name goods. Another change is on the way: the Chinese are moving to Brooklyn, Queens, and other suburbs.

Toronto, Los Angeles, and almost every city in Canada and the United States have their share of developmental controversies, pitching Blacks against Whites, immigrants against natives, or Evangelicals against Catholics. It could be a seniors' residence for Italians, a Korean strip mall, an Islamic mosque, or Chinese store signs. I will have more to say about some of these controversies in chapter 10. Here I am concerned with observing the patterns of such controversies. These are strikingly similar in various cities.

Storefront signs in non-English languages raise the question of openness to others. The local councillor of Flushing wanted Chinese and Korean stores to have English along with their own languages on their storefront signs. Interestingly, he was himself a Chinese American of forty years' standing who owned a chain of drug stores. Yet there was a widespread reaction from the store owners, who viewed the suggestion as an attack on their rights as Americans.[40]

Seven miles from downtown Los Angeles is the city of Monterey Park, which has become the first majority Chinese ethnoburb of the United States. By the estimate of the local Chamber of Commerce, two-thirds of its 5000 businesses are Chinese-owned. As part of the politics of representation and control between newly arriving Chinese and long-established White native residents in the 1980s and 1990s, Chinese storefront signs became a lightning rod. The issue roused the English-only movement, which succeeded in 1986 in getting a city council resolution passed to that effect. Though purely a symbolic success, it mobilized the Chinese community to elect their own representatives for city council in 1988. That practically killed the movement. Chinese signs now march along major roads over long stretches of the valley, spreading beyond Monterey Park.

These communal controversies vary in their local configurations. Yet their scripts come from the same book. The competition for space, jobs, services, or symbolic representation between groups of distinct identities builds up mutual wariness. A proposal for some change ignites distrust that begins to be reflected in civic institutions, public decision-making processes, and the media. Hard-line leaders emerge on both sides of the dispute. Confrontations are virulent and charges of racism fly. The controversy is acted out largely in public institutions. Some losses and gains are traded. Eventually equilibrium is reached through legal and political interventions. Over time the context begins to change and broader social changes swamp contentious feelings. The controversy fades away.

There is a distinct class of social controversies arising from the introduction of (new) religious symbols and practices in the public realm. They precipitate a clash of values in the public realm.

Religion in the Public Realm

The place of religion in the public sphere is a neglected topic in the discourse on multiculturalism. The secular bias in the conceptions of multiculturalism forecloses any discussion of religion in the public sphere. Yet religion is woven into the cultures of the mainstream in the form of Judeo-Christian traditions. Immigrants are bringing new, non-Judeo-Christian religions, into North American societies. Their practices and beliefs enacted in the public sphere become a source of communal controversies. New citizens' assertions of freedom to practise religion clash with secular demands that the public sphere be free from religious expressions. Public controversies rage about demands for the presence in public places of Sikhs' turbans and kirpans (ceremonial daggers worn by religious command), Muslim women's face-veils, prayer rooms in schools and offices, the orthodox Jewish custom of erecting Sukkahs (temporary shelters built for prayers for seven days of the Sukkot festival) on balconies and in backyards, and Islamic and Hasidic Jews' emphasis on the segregation of women.[41] Some incident of a Sikh student wearing kirpan in a school, a woman insisting on keeping her face-veil on in a court, or a group holding prayers in a public place blows up into a public controversy, with supporters and opponents lining up claiming public values are on their side. The media amplifies such controversies fuelling extended public discussions about the expression of religion in the public sphere.

The arguments are legalistic, political, and cultural. What is significant from the viewpoint of living in multicultural cities are the sentiments and perceptions aroused by these controversies. Groups whose practices or beliefs are the object of controversy feel defensive and picked on, while those on the other side perceive a threat to existing norms and values. Some of the controversies end up in court, and others are resolved through community compromises, while some others simmer for a while and then cool down without any material change. Yet all of these feed into the image of different communities, which in turn define mutual perceptions. Only rarely do such controversies create a level of communal disharmony that spoils everyday relations.

Sharing the City

A multi-cultural city not only affirms cultural differences but also infuses a spirit of unity among citizens. It has a civic culture of shared symbols, experiences, places, and spectacles for all segments of society.[42] Civic culture arches over cultural differences. It draws people out of their communities and groups into the common ground of shared sentiments, feelings, and identification. Public places and activities are the framers of civic culture. These are the sites where strangers come to bask in one another's company and share a common vision of social life. What are those places and activities?

Not everyone goes to the baseball games of the Blue Jays (Toronto), Mets and Yankees (New York), or Dodgers (Los Angeles), but the local team's pursuit of a championship thrills people across ethnic and social lines. Mark Hutter observes that baseball (also other sports) has been a "form of urban landmark that drew people together and became a symbolic representative of the city as a whole."[43] Public places and events provide "a vision of social life in the city" for those who live there, tourists who visit, and commuters who come for work, shopping, or entertainment.[44]

The landmarks of a city symbolize its spirit and identity. They bring people of all backgrounds together around common visions. Toronto's

CN Tower and Harbourfront, New York's Empire State Building, Statue of Liberty, and Fifth Avenue, Los Angeles's Bunker Hill and Bankers' Tower, and Hollywood's hill-side sign are some of the symbols that join residents in a common imagery. Regardless of race, ethnicity, or class, a resident takes a visitor to these places to "present" the city. In today's corporatized world, the image of a city is increasingly being influenced by giant electronic billboards, the sights and sounds of commercial streets, mega malls, and recently by signature architecture. Designed by Frank Gehry, Los Angeles's Walt Disney Concert Hall, New York's Beekman Tower, and Toronto's Art Gallery of Ontario are symbols of the new urban aesthetics. These places are on view for all to see and identify with, even if they are not entered. Similarly, museums, zoos, theatres, cinemas, and music halls are symbolic unifiers of citizens, even if not all benefit from them. They are a form of optional consumption shared by residents of a city.

Other elements of civic culture that pull people together are parades, fairs, and summer festivals. As well, events such as marches, demonstrations, or protests can evoke common passions. The riot sparked by the G-20 heads of state meeting in Toronto (2010) and subsequent arrests was on the 24-hour news cycle and absorbed almost everybody for a few days. The Occupy Wall Street movement's encampment in Zuccotti Park, New York (2011) was the talk of the town for weeks. Obviously, these happenings do not result in consensual visions, but they do bring a common focus, however temporary, to people's attention across the social spectrum. The same is the case with news of sensational crimes and political scandals. The good and bad are both elements of civic culture that raise the sense of a conjoined fate.

Civic culture is not just a representation of mainstream interests. Elements of ethnic cultures come to form part of civic culture and acquire the status of common practices and symbols. Of course, the power of a group affects the integration of its symbols and practices in civic culture. Yet the economic appeal of cultural diversity, such as its usefulness for attracting tourism and global trade, opens civic culture to ethnic influences. For example, the Caribbean Festival in Toronto features calypso, steel bands, and a flamboyant costume parade. It draws about a million people and unleashes a week of festivities in the city. Not as large but equally striking is the annual West Indian Carnival in Brooklyn, New York. These are examples of public events that have grown beyond their ethnic niches, as has been the case for the St Patrick's Day parade. They function as unifying instruments of civic culture.

Mainstream Culture as the Carrier of Diversity

A theme that runs through the above discussion is the role of the laws, ethics, values, and norms of mainstream society, both national and their civic (city) representation, as the baseline for linking together ethnic subcultures. A common language, English in North America, is also necessary for living in the same space. The everyday interactions among persons of diverse cultures are negotiated ultimately by resorting to common laws, regulations, ethics, and trust. Whether they concern racist police practices, Muslim women wearing veils in public places, or Chinese building mega-homes in historic districts, all such ethnic controversies are ultimately arbitrated on the basis of laws, regulations, and ethics as they evolve on the main street. In the literature of multiculturalism, the role of the common ground of historical but changing main-street culture is not recognized. It is overlooked in the zeal to advance the right of cultural differences. Of course, the main-street culture continually evolves, incorporating ethnic values and practices, as have been argued in this book, particularly in the above examples of everyday multiculturalism. Yet it is functionally necessary that in its evolution it serves as the mediator among competing values. Without reaffirming the role of a common ground, the multicultural city will not function. It will be a dysfunctional, fragmented city.

Conclusion

Multicultural cities are exciting places to live. The exposure to a wide variety of people, languages, and cultures makes living in these places an adventure of daily surprises. The variety of stimuli and choices heightens one's sensibilities, particularly as these exposures and encounters demand respect for the right of others to be different. Yet the variety and differences are counter-balanced by forces of unification lodged in civic culture and the city's institutions. This dialectic of diversity and unity leads to changes in the perceptions and practices of ethno-cultural groups, on the one hand, and the norms and values of the overarching civic culture, on the other. This process of dialectical change is particularly evident with food, music, etiquette, sports, and art.

Everyday life at the individual level is marked largely by tolerance and civility, in other words, by an attitude of indifference to differences. But there are tensions of diversity arising from differences in race, culture, and even values. Inter-group relations are the main arena for tensions sparked by some perceived inequity, narratives of distrust, or political competition for power and control. The experiences of group-level interactions are at variance with people's dealings as individuals. This divergence is reflected in narratives and imagery. A common ground of shared laws and citizenship ethics is necessary for overcoming such differences.

The multiculturalism of the city is experienced from the base of one's ethnic or social moorings. In food, music, media, literary and religious associations, parades, fairs, and social networks, different groups have parallel public lives. Yet the city draws people out daily to encounter, deal with, and relate to others through other public institutions. This is what makes for the "feel" of living in a multicultural city.

Chapter Eight

Political Incorporation and Diversity

Diversity, Equality, and Local Politics

On a sunny day in June 2011, a coalition of Asian New Yorkers demonstrated outside city hall demanding public recognition, political representation, and a fair share in the services and grants for what they claimed to be a one-million-strong Asian group, comprising forty nationality communities.[1] The *New York Times*'s headline captures the essence of the politics of racial and cultural diversity: "Asian New Yorkers Seek Power to Match Numbers."[2]

Ethno-racial diversity when combined with equality rights turns into the driving force for demands of recognition, representation, and a fair share, which in turn become instruments of infusing multiculturalism in a city's political and administrative institutions. How does this transformation happen and what forms does it take? In other words, what changes occur in the configuration of political power and processes of policymaking in response to social diversity? This question will be answered by examining the processes and patterns of urban politics and policymaking in three acknowledged multicultural cities, Toronto, New York, and Los Angeles.

The demands for recognition, representation, and a fair share are articulated in the context of a specific space and time. Even if the demands are national in scope, it is in the local context where they are articulated and raised. Dr Martin Luther King Jr's revolutionary "I have a dream" speech (1963) was appropriately made in Washington, DC, the national capital. Cities are the venues where immigrants mostly settle, even though their arrival may have been orchestrated by national policies. Therefore, city politics and policies are critical for studying the multicultural transformations of political institutions.

Of course, local governments and communities are not sovereign states, self-contained and masters of their destinies. They are the creation of higher-level national, state, or provincial governments. They may have a fair degree of autonomy under "home rule" or city charters, or, alternatively, be dependent on higher levels of government, but one way or another they remain the vassals of the national and provincial authorities. Similarly, urban economies, political organizations, and social structures are patches in the fabric of national, sometimes global, societies.

North American cities though differing in authority and capability have many comparable conditions and institutions. They are responsible to varying degrees, depending on the constitutions of local governments, for providing utilities, infrastructure, transportation, land and environmental management, welfare and health services, urban planning, policies for housing, local economic development, provision of education, child and elder care, and many other activities necessary for the health, welfare, and prosperity of an area and its population.[3] While many of these are collective goods whose production and distribution may be contracted out, they are developed and financed in the public interest.

People's needs for jobs, housing, services, and social rights are met locally in the first instance. Peter Drier, John Mollenkopf, and Todd Swanstrom argue that place matters in determining people's quality of life and sense of fair play.[4] The point of this argument is that locality and its politics and policies are critical instruments for fulfilling individual and community needs. What are the patterns and processes for articulating, accommodating, and incorporating ethno-racial communities' demands? This question is to be explored presently.

Demands as Policies

Ethno-racial groups' demands for equity in the public sphere are often articulated as desired policies and programs. Political arguments are about an equitable share in jobs, business contracts, housing, quality of schools, police protection, or the culturally sensitive provision of community facilities and services, and not just about recognition or representation in public institutions. Even the struggle for political power is often motivated by specific grievances and needs. Police profiling of non-Whites or the need for bilingual education helps mobilize ethno-racial groups to strive for political influence, and energizes communities to vote for their own representation in councils of authority. This is how the process of political participation unfolds. It does raise expectations of a fair share in political representation, but it evolves from the roots of concrete demands.

The responses to the demands of ethno-racial communities from both public authorities and corporate organizations tend to follow a defined set of strategies and practices. Kristin Good has developed indicators to measure the responsiveness of Canadian municipalities to multicultural needs.[5] She uses indicators such as creating an inclusive municipal image, involving minorities in policy discussions, increasing access and equity in service delivery, introducing employment equity, instituting multicultural and anti-racism initiatives, offering community grants, promoting multicultural festivals and events, and developing immigrant and settlement policies.[6] These indicators essentially reflect policies for the integration of ethno-racial minorities. Good's list of policies leans towards the recognition aspect of political incorporation, with some demands for equity in services. Representativeness and a corresponding sharing of power are not high on her list. American theorists lay primary emphasis on minorities' representation and share in political power.[7] Building on the foundation of Good's list and adding goals referring to political and administrative representativeness and an equitable share in services and resources, the following list of policies is offered as a comprehensive approach to minorities' inclusion.[8]

Recognition
- Creating an inclusive municipal image, through symbols and public statements.
- Establishing participatory and consultative arrangements for involving ethno-racial communities in policy deliberations, including language support and interpretation services.
- Supporting minorities' and immigrants' community organizations and promoting their involvement in local affairs.
- Supporting multicultural festivals, nationality parades, and other public expressions of diversity.

Representation
- Opening of political parties to minorities, promoting voter registration, supporting community mobilization, and making electoral politics more inclusive.

- Promoting policies and practices for increasing ethno-racial groups' share in the elected and appointed offices of local authorities, boards, and committees.
- Increasing the representation of diverse groups in local administration through employment equity and diversity recruitment.
- Establishing anti-racism and diversity norms as corporate policies.
- Introducing diversity, cultural, and racial awareness training in public and private organizations.
- Promoting and monitoring minorities' and immigrants' representation in local government and community organizations.

Equity provisions
- Increasing ethno-racial communities' access to housing, child and elder care, health, recreation, education, security and police, justice, culture, art, and other services, particularly at the local level.
- Accommodating the cultural and religious mores of ethno-racial groups in local programs, services, and practices, making provisions for the culturally sensitive delivery of services.
- Enacting equal-opportunity provisions in businesses, jobs, and the distribution of resources. Removing racial and cultural barriers to equal opportunities in the local economy and environment.
- Providing a fair share of grants to minorities' community organizations and cultural associations.
- Providing settlement services for immigrants and special programs for disadvantaged minorities.

These policies are statements of how demands for recognition, representation, and a fair share in the governance of a city can be concretized. The responsiveness to diversity can be judged by the incidence and enforcement of these policies. Yet seldom will every multicultural city enact all these policies. Often cities differ in their political and administrative systems and resources. Also, local ethno-racial groups' priorities differ and they pursue demands that are pressing in their respective circumstances.

The Politics of Inclusion

The means to realize these policy goals are primarily political. It is organized political activity that opens paths for all groups to participate in policymaking and be included in the exercise of power. Minority groups in particular have to engage in public discussions, media campaigns, lobbying, protests, mobilization of their voting power, and seeking elected as well as appointed office to have their priorities incorporated in social agendas. Generally, voters are reluctant to support the provision of public goods if they perceive that there is an ethnic gap between them and the beneficiaries of those services. Even if one has some reservations about the broad sweep of this observation, it is self-evident that political incorporation is a necessary condition for communities to realize their demands.

Viewed comprehensively, political incorporation is the process by which a self-identified group articulates, expresses, and represents its interests and stakes a claim in policymaking.[9] In practice, it means participating in and influencing decision-making processes to obtain a fair share of public and private goods. Recognition, representation, and a fair share of resources are the outcomes of political incorporation.

There are two distinct sets of factors that affect the process of political incorporation: external or contextual to a group and internal or strategic within its field of choice. They correspond to a structure versus agency dichotomy. This duality is a useful perspective for discussing the conditions for and ways of achieving political incorporation.

Among the structural factors affecting the political incorporation of a group are eligibility rules (residency requirements, citizenship laws, etc.), voter registration regulations and practices, the organizational practices of political parties, a group's geographic distribution (residential concentration vs. dispersion), the demarcation of electoral districts (gerrymandering), and the local power structure, economic organization, and political culture.[10] It is a long list, to which many other items can be added depending on the differences among cities and countries. The point is that there are many environmental or structural variables affecting the political incorporation of diverse groups.

Agency factors are based on a group's capabilities and resources for mobilizing and organizing itself to manoeuvre towards its goals. What a group does to gain representation and influence in decision making is an expression of its capacities for agency. Among the strategies commonly used by identity groups are appeals to national values and ideologies, voter registration and mobilization, media campaigns, legal challenges, marches and protests, joining political parties and aligning with mainstream political brokers, coalition forming with sympathetic groups, and, in extreme situations, riots. Of these measures, only some are relevant to a situation, depending on local conditions and the resources of a group. Of course, they presume pursuing political goals by democratic and legal means in a political system. There are various levels of radical and revolutionary discourses, which in North American cities haven't had much track.

A basic point to be underlined is that the historically evolved political system and constitutional order lay down the paths for access to political power and influence. Ethnic groups and immigrants mostly follow the paths thus laid out. They have to work through the medium of the mainstream political system, with some incremental changes. It is rare for any group to carve out a radically new path to political power.

Ethno-racial Politics in Multicultural Cities: New York

New York City has a centralized system of governance. Its directly elected mayor has sweeping powers, backed by the city council, and five borough presidents, who have largely symbolic powers.[11] It has fifty-one ward-based councillors, a fact that allows minorities from ethnic neighbourhoods to capture council seats. Its political system is organized around political clubs, business and labour organizations, and community groups. With a large number of appointed and elected offices on community, school, and city boards as well as commissions, ethno-racial and various interest groups have many opportunities to wield power, even without winning the big prize of the mayoralty.

Historically in New York, ethnic communities have followed the strategy of forming coalitions with labour and community organizations and aligning with political parties, mostly the Democratic-liberal ones, by promising them their votes in return for political influence. Tammany Hall is the metaphor for this form of politics through which generations of immigrants, particularly Irish, found political incorporation. Italians and Jews and, in the post–civil rights era, American Blacks have resorted to similar strategies. Now Dominicans, Puerto Ricans, Latinos, and the Chinese are following more or less the same script.

Ethnic politics operates within the ideological framework of the conservative-liberal divide, broadly represented by the Republican and Democratic parties, with a periodic flaring up of third parties and the rise of radical groups, as in the heady days of the anti–Vietnam War and Black Power movements of 1960–70s. Yet these parties only endorse candidates, and are not represented on the city council. Coalition formation between minorities and established ethnic groups has been a strategy of gaining access to political influence. A historical example is the alliance between liberal Jews and Blacks in the 1960s. John Mollenkopf and Raphael Sononshein observe that "the ethnic and racial politics confronting today's immigrants to the great U.S. cities are not blank slates. They have been already shaped by waves of previous newcomers who have established enduring patterns of intergroup conflict and alliances."[12]

Ethno-racial groups differ in their success in gaining political influence and meeting their needs. Some do better than others and those ascendant in one period may fade away in the next round. Racial minorities, immigrants, and the poor have continued to lag in the fulfilment of their community needs. And the gains they make, say in housing or school quality in economically expansive and politically liberal times, are reduced by shifts in the political cycle towards conservatism and in periods of fiscal restraint. Towards the end of the first decade of the twenty-first century, recession and a budgetary crisis combined with the rise in the influence of the Tea Party Republicans has shredded the social support networks built up over the past many decades in US cities.

All in all, the political incorporation of ethno-racial groups proceeds within the workings of political and economic cycles. It seldom leads to the complete equalization of political power, but it has led to the increasing representation and influence of minorities in public policymaking.

In 1980, Whites formed 52% of the population in New York and held 74% of the city council seats. Their numbers decreased to 35% of the population and their representation in council fell to 51% by 2000. By 2012, Whites, though still dominating, had 47% of the council seats, with only about 33% of the city's population being non-Hispanic White.[13] The population of Blacks (both native and foreign born) increased from 24% in 1980 to 26% in 2010, while their percentage of councillors increased from 17% to 28%, reaching demographic proportionality in representation. The Hispanic population increased from 20% in 1980 to 29% in 2010; correspondingly, their council representation increased from 9% to 20%, again a notable increase but below the percentage of their population. Asians had no representation on city council in 1980, but they had two seats, making up 4% of the councillors in 2012, while their population increased from 3% to 13%.[14] The representation of ethno-racial groups has steadily increased in the last thirty years. The ethnic diversity of the city council is reflected in its councillors' profile: it had three West Indians, two Dominicans, one Chinese, nine Puerto Ricans, and ten African Americans in 2009.[15]

These figures show that minorities have made gains in political representation since 1980. Historic barriers to political offices have been breached, though new minorities still do not exercise political influence proportionate to their population.

It must be pointed out that ethno-racial representation is one of the many political processes in a city. It operates alongside the drives for power by social classes, labour, business interests, women, and gays. Also, neighbourhoods' political mobilization for services and development bring about alliances that cut across ethno-racial loyalties. Within the scope of its liberal values, New York's political culture is aligned along a liberal-conservative divide. It has had only one non-White mayor, David Dinkins from 1990 to 1993, followed by eight years of conservative White Republican rule under Rudolph Giuliani, who rolled back many services for minorities and the poor. Michael Bloomberg, of Republican leanings and a billionaire, dominated city politics for a decade, steering the city through the post-9/11 setbacks, Hurricane Sandy's devastations, and the economic recession of 2008. Yet he promoted diversity and advanced immigrants' inclusion.

The strategies for political influence employed by ethno-racial groups, gays, and women are beginning to show new configurations of coalitions, intergroup competition, and leadership patterns. West Indians, and to some extent Dominicans, are adapting to the opportunities offered by their increasing numbers to build strong political constituencies, while American Blacks and Puerto Ricans are under pressure, though Blacks have a long history of agitating for equity.

Dominicans, as the largest immigrant group in New York, have not gained a proportionate share of city-wide political offices, but they have captured a district Democratic Party leadership. They control the Washington Heights Area Community Planning Board and have replaced Jews on the district's six school boards.[16] Other recent immigrants such as Indians, Pakistanis, and Chinese are initially focused on building their cultural and religious institutions. Homeland politics also preoccupies these groups, as many national political parties have sponsored their overseas branches and the satellite TV keeps immigrants continually in touch with the politics back home. This inward focus of their attention may be inhibiting some groups in striving forcefully for political participation in the city. Yet they have gained many appointed positions and are well represented in the professions and media.

Overall, one can conclude that most ethno-racial groups have not attained a proportional share of political influence and some inequalities persist. Yet the political incorporation of these groups is proceeding and most have gained influence if not power through the mobilization of their demographic, cultural, and economic resources. These groups are not facing an impregnable power monopoly. There are non-electoral routes to political influence (such as a media presence and the formation of public opinion, professional advancement, and community organization), which will be discussed later. I will also discuss later public recognition and equity as elements of political incorporation.

Ethno-racial Politics in Los Angeles

Los Angeles city is one of the eighty-eight municipalities within Los Angeles County, which is itself a part of the large megalopolitan region called the Los Angeles–Long Beach–Riverside Statistical Area. Politically, the city and county divide up responsibilities for local services. The city is largely responsible for property-related services, such as police, fire, zoning, and planning, and recently it has acquired control over schools. The county provides social, health, and welfare services, and many autonomous boards and commissions deliver water, power, transportation, and other services. Overall, it is a fragmented political system, with competing centres of power and authority.

Los Angeles city council has only fifteen members, with the mayor as the executive head. Its electoral districts are large and spread out, making it difficult for any one ethno-racial group to carry a councillor's election by itself.[17] Furthermore, there are large residential concentrations of Blacks, Latinos, and Asians extending across the city boundaries into the suburbs of the county, such as in South Central Los Angeles, which includes a million Black residents spilling into the county. Similarly, a large sector of Latino concentration extends from the downtown to East Los Angeles. Obviously, these districts hold out promise for Blacks and Latinos to elect someone of their own.

Compared to New York, Los Angeles city operates largely in a nonpartisan mode, wherein even the most popular party, the Democrats, is weak at both the city and county levels.[18] Local party clubs are not the primary path to political power, though they have been stepping stones for many. Historically, the chamber of commerce and labour unions collaborated in running the city. The suburbs had local political organizations, often guided by the development interests that dominated electoral politics.

The Los Angeles area has a high percentage of recent immigrants, for instance Latinos, who are not citizens and thus are not eligible to vote. Also within the ethno-racial groups, there is a predominance of some nationalities, such as Mexicans among Latinos and Taiwanese among Chinese. A few large groups vying for political influence have an incentive to going it alone and avoiding forming coalitions. These demographic factors, combined with the differences in the Los Angeles's governmental system, create a structural environment that is different than that of New York.

Before the civil rights era, American Blacks, though excluded from voting, were organized by Black churches, political clubs, and the National Association for the Advancement of Colored People (NAACP). During the 1960s, the Black Power movement gave a revolutionary edge to their struggle. Black leaders ran successfully for the city of Los Angeles council after the Voting Rights Act (1965).[19] Their strategy of aligning with Jews and Anglo liberals, mostly Democrats, paid off in the election of the first black mayor of the city, Tom Bradley. He was re-elected five times, and served for twenty years (1973–93) as the mayor of Los Angeles. This marks a major difference in the political cultures of Los Angeles and New York

Of course, the swing of the political pendulum prompted by the 1992 Rodney King riots generated resulted in the election of a White Republican, Richard Riordan, as the mayor. It mirrored the political process in New York after the tenure of Black mayor David Dinkins. Now the Black population is beginning to decline proportionately and immigrant groups, Latino in particular, are on the rise politically as the largest ethno-racial group.

In 2005, Antonio Villaraigosa, a Mexican American, put together another Democrats' coalition of Blacks, Latinos, and liberal Whites to win two successive terms in the mayor's office. In the process, Mexicans broke through the political power structure in the city.

The city council had started to reflect the political incorporation of ethno-racial minorities earlier. In 1990 non-Hispanic Whites, who formed about 37% of the city population, had the majority of elected councillors, 60% (nine out of fifteen council members), Blacks had three representatives (20%), with a population base of 14%.

Latinos remained underrepresented, having 40% of the city's population in 1990 and 13% (two councillors) of council members. Asians had one (7%) councillor, although they formed 10% of the population. In 2012, out of fifteen councillors, four were Mexican Americans, another of mixed Mexican ancestry, with three African Americans and no Asians; almost 47% of the councillors were from minorities.[20] In 2010 Los Angeles County's board of supervisors had only five elected members, including the mayor. Out of the five one was Black, one Mexican, and three were White, including one of Jewish and another of Croatian heritage.[21] Many Mexicans and Asians have been elected in the eighty-seven suburban cities, towns and villages of the county. For example, in the city of Monterey Park, a predominantly Chinese suburb, four out of the five councillors are Asians, including the mayor. This change has occurred since the 1990s with the settlement of Asians.

Undoubtedly, ethno-racial groups, bolstered by immigration, have been gaining representation in the political power structure, moving along the paths charted primarily by earlier immigrants, such as Jews, Italians, and Irish, and more recently by Blacks, through the civil rights struggles. Yet their electoral representation rises and falls with changing demographic and political-economic conditions. Obviously ethno-racial politics are buffeted by broader political currents and community movements.

Among the strategies minorities have followed are the old formulae of coalition formation, aligning with ideologically close mainstream factions, joining political parties, voter registration, and community mobilization, taking advantage of their residential concentrations and nationality solidarity. Tapping the energy generated by radical movements and agitations has also been used by leaders of minorities to advance their march to power. These strategies also usher them into public-interest politics as they gain responsibility for decision making for cities as a whole. Minorities' leadership politics is joined with the ideological streams of the mainstream political parties. It becomes a process of mutual learning and the forging of a common ground. How do these patterns compare with the politics in the Toronto area?

The Politics of Diversity in the Toronto Area

The politics of inclusion works differently in Canada, where the emphasis is largely on how political institutions respond to the needs of ethno-racial groups and immigrants. Political incorporation is assessed in institutional rather than representational outputs. Kristin Good's approach of developing a framework to measure the responsiveness of municipalities to immigrants needs is an example of the institutional lens through which political incorporation of ethno-racial groups is viewed by many Canadian analysts.[22] Myer Siemiatycki, Caroline Andrew, John Biles, and other scholars of Canadian immigrants' political integration acknowledge that this is a neglected topic in the literature.[23]

The political cultures of Canada and the United States differ. In the United States, political representation is an end in itself, and race and ethnicity have been historically the basis of the distribution of political power. Canadian power dynamics has been dominated by its English-French divide, with other ethno-racial groups historically left on the margins. Canada's Charter of Rights and Freedoms, multiculturalism policy and act, as well as the political imperative of courting ethno-racial votes have induced political parties to include minorities. In Canada, representation is bestowed by political parties rather than taken by ethno-racial groups through mobilization. These differences of political culture have created some very interesting but contrasting processes in the two countries.

Canada has had two successive visible-minority (of Chinese and Haitian origins) governors general, the vice-regal representatives of the British monarch, as the symbolic head of the Canadian state, from 1999 to 2010. These two non-White heads of state obtained their positions by appointment by the Liberal party[24] much before the United States elected its first Black president in 2008. Yet Barack Obama's election has been seen as a historic, breakthrough event for non-Whites, because it symbolized Blacks' crowning success in gaining political representation. This example highlights the differences in the political cultures of the two countries.

Other structural features affecting the process of political incorporation in Canadian cities are the circumscribed role of cities within the provinces, the division of local authority between city and higher-tier (regional) municipalities, the non-party basis of local elections, the cities' limited scope for taxation and borrowing, high rates of citizenship among immigrants and their growing voting power, a historically small non-White population, and a relatively non-litigious culture.

Toronto is the nucleus of the Greater Toronto Area (GTA), which includes the city, four second-tier regional municipalities, and twenty-eight local or first-tier municipalities.[25] In the chapter 4, I discussed the demography and social geography of the ethnic enclaves that extend out from Toronto city to the suburban municipalities, such as Markham, Mississauga, and Brampton. The existence of ethnic enclaves could favour the election of ethno-racial representatives, but interestingly the results are not striking in this regard. Low voter registration and poor turnout of the newly naturalized citizens at elections have contributed to the election of just a small number of ethno-racial representatives, much lower than the percentage of these communities' populations.

The consolidation of four suburban municipalities with the city of Toronto (1998) resulted in the drastic reduction of municipal councillors from 106 to 58, which was subsequently reduced to 45 in 2000. This change expanded the size of electoral districts and thereby increases the probability of their becoming multi-ethnic wards. Yet the patterns of ethno-racial distribution of political representation have changed little.

The Anglo-Canadian ancestry group formed 30.3% of the city's population in 2001, but had 53.8% of the pre-merger elected representatives, a percentage that fell to 44.4% in the 2000 elections. Other ethno-racial groups who consistently have had more than their proportionate share of representatives include Italians and Jews. In contrast, non-Whites formed about 42% of the population in 2001, but had 11.1% of the council seats in 2000.[26] By 2011, the non-White (visible minority) groups, who are the new face of ethno-racial communities, formed 49% of the population, but were represented by just five city councillors (11.1%).[27] There has been no change since 2000.

The representation of visible minorities in the GTA was not much better. Out of the total of 253 municipal councillors in 28 municipalities in 2011, only 18 (7.1%) came from visible minorities, while they formed 47% of the population.[28] At the provincial and federal levels, the picture is slightly different, though visible minorities were still underrepresented.

There is some indication that ethnic voters are not swayed by the ethnic loyalty only.[29] In the Toronto area, the results of the 2014 local elections suggest that Chinese and South Asians did not elect their compatriots even when there were numerous candidates of such backgrounds in the ethnic enclaves and wards. They seemed to have voted on the basis of the credibility of candidates in terms of understanding and advancing civic interests. The non-ethnic incumbents who had served them well won handily, even in wards of ethnic concentrations.

In 2004, Toronto city elected members of visible-minority background in the following percentages: three members of the provincial parliament (14% of Toronto-area MPPs) and two members of the federal parliament (9% of Toronto MPs).[30] In the 2011 federal and provincial elections, out of the GTA's 47 provincial legislators, 12 (25.5%) came from visible minorities, whereas their share on the basis of their population would have been 19. Among 47 federal members of parliament from the GTA in 2011, only 8 (17.1%) were of visible-minority background.[31]

Overall, the Toronto area shows a pattern of a relatively higher level of ethno-racial representation at the provincial level, followed by the federal parliament, and the local councils, the least representative. Irene Bloemraad observes a similar trend in Vancouver: "We again find a greater degree of diversity at the federal level, where two of the five MPs were visible minorities, compared to three of ten MLAs (provincial assembly members) and only one out of eleven members of city council."[32] Yet at all three levels, by the proportionality criterion, new ethno-racial groups were underrepresented, with little change over time.

Federal and provincial political parties court the minority vote by opening their doors to ethno-racial groups. Enterprising and politically savvy individuals from these groups recruit their compatriots to pack the party nomination meetings for federal and provincial constituencies. The parties extend recognition to such nominees as their multicultural face, allowing their nomination to stand, and thus they get elected on party tickets. For example, Sikhs, a relatively small South Asian group (less than 1 per cent of the Canadian population in 2001) held six federal seats and eight provincial ones, largely in the Toronto and Vancouver regions, where they are concentrated.[33] Their electoral success came from their community mobilization, political awareness, and use of the constituency political party as a ladder. At the municipal level, elections are on a non-party basis and minority candidates have to compete on their own. They are thus not as successful as in provincial and federal elections. Also, incumbency gives an advantage to an elected representative of both name recognition and a reputation built by serving the electorate.

It is also possible that ethnic politics has a limited traction among minority groups. They discover that joining mainstream politics offers more rewards than concentrating on electing ethnic representatives who may not have enough political clout to meet their needs or an understanding of public issues. They quickly learn to be invested in mainstream politics, seeking political influence through the established political paths. For this reason, they pursue provincial and federal parliamentary offices, presumably to have more influence and prestige.

This, in broad outline, is the process by which higher-level governments tend to have proportionately greater minority representation than local councils. It is not to deny the commitment to equity and pluralism of the parties or the political ambitions and mobilization of minorities. Yet the earlier point that Canadian political culture grants representation to ethno-racial groups on its own terms is borne out by the above description of the patterns of ethno-racial representation. Of course, it bears reiterating that minorities' needs are also responded to in pursuit of multiculturalism policy and equity goals.

At this juncture, another question arises, namely, what benefits does representation bring? Is a share in power necessary for meeting minorities' needs?

Does Representation Matter?

An iconic image of question hour in the Canadian parliament, often seen on TV, is the prime minister standing up and speaking with, positioned prominently behind him, a Sikh MP of the ruling party in his colourful turban, clapping and smiling. Political rumour has it that some Sikhs MPs have not said a word in parliamentary debates in the four years of their tenures. Whether this is true or not, it highlights the question of whether electing representatives is enough to bring about the political incorporation of ethno-racial groups?

Representation that does not lead to any significant input of ethno-racial interests into policymaking is often described as tokenism or co-optation. Yet minorities' interests being taken care of by the political power structure without their representation is itself a form of paternalism. Neither condition is conducive to a sense of belonging and integration.

Representation is, simultaneously, an end in itself and a means to the ends of political equality and incorporation. As an end, it has both symbolic and substantive values. Symbolically, the presence of minority, women, and gay representatives signals the legitimacy of those groups' stake in the power structure. It concretizes equality and breaks down discriminatory barriers. Regardless of their contributions to legislative debates, Sikh MPs in the federal parliament communicate to millions by their presence that Sikhs are both full citizens and custodians of Canadian national interests. They symbolize political integration.

Representation brings benefits to the represented group through the recognition of its needs, and the opportunity to bargain for political rewards for communities, such as programs, contracts, and jobs. Political representation is the highest form of employment equity that sets an example for the society as a whole.

Theoretically, there are two models of representation. In a collectiv-ist conception representatives are essentially agents of particular social groups, with the implication that public interest is a composite of sec-tional interests.[34] By contrast, the individualistic representation means a representative is a free agent acting on his or her understanding and beliefs. This is a liberal individualist view of political representation, wherein a representative, once elected, acts according to his/her con-science and values for the common good. In practice, politicians act on the basis of individual values and beliefs, but they also are influenced by groups who contribute to their election, be those political parties, ethno-racial communities, or ideological or financial supporters.

Ethno-racial representatives have to act within their parties' man-dates. They have to learn to be public representatives of all communi-ties in their constituencies and not just be focused on their own group's demands. The public interest becomes the touchstone of their political behaviour. Thus, the common ground of the mainstream political cul-ture bears down on them to work for the collective welfare while seek-ing benefits for their communities.

High hopes projected onto ethno-racial representatives are seldom fully realized. In political decision making demands are modified, ide-ologies are compromised, and outcomes are moderated to negotiate with other interests. The success of the representation of ethno-racial interests should be assessed in the context of the scope of the political process.

Myer Siemiatycki and Anver Saloojee analysed the relationship between the ethno-racial identity of councillors in the city of Toronto and their political stance on various issues of community concern.[35] They conclude that the voting record of visible-minority councillors on social issues that were of concern to their communities was "more conservative than progressive" and "there is no necessary correla-tion between a politician's identity and his or her stance on (these) issues."[36] They also cite studies that echo similar findings about the "ambiguous impact of women elected to the federal parliament" in Canada.[37] Furthermore, they document the responsiveness of Toronto city towards immigrants' concerns enacted by a group of White pro-gressive councillors. Irene Bloemraad observes about Vancouver city that "despite the lack of mirror (proportionate) representation – the city of Vancouver can boast of many successes."[38] Responsiveness to minorities' needs has a weak positive correlation with their political representation in Canada.

The political representation of minorities and immigrants in the US cities has a different meaning. It is popularly regarded as a necessary condition for political incorporation. The value put on electing minorities' representatives is very high, because of the long history of keeping out Blacks from the power structure. The representation is often not proportional to various groups' population, but electing persons of a particular identity is a mark of political and social equality for them. Representation is valued for its own sake.

The relationship between representation and responsiveness to minorities' concerns in the US cities is also weak. New York has a less than proportional representation of immigrant minorities. Blacks are approaching proportionality after years of struggle. Yet the city measures well on indicators of recognition and equality of service delivery, despite periodic accusations of racism in schools, police, and neighbourhood services.[39] In a study of four small New England cities, Peter Burns found the responsiveness to Black and Latino interests varied almost independently of their representation. Structural factors such as local political parties, non-electoral pressure groups, economic power structures, and minority leadership affected the responsiveness.[40]

Another dimension of ethno-racial representation that has received little attention in the literature is that there cannot be proportional representation of all ethno-racial groups. There are too many to be represented. Analysts tend to overlook the wide diversity of race, ethnicity, nationality, and religion within broad categories, such as non-Whites, Asians, and Latinos. An illustration of this point is the case of the Toronto area's non-White representation discussed above. Among the elected representatives, Sikhs dominate among South Asians and Chinese among other Asians, leaving Filipinos, Japanese, Afghans, Pakistanis, Somalis, Russians, and hundreds of other identity groups without representation. Obviously, municipal councils or provincial parliaments are not assemblies of communal interests. They are unifying bodies looking after common interests with equity for various groups.

To conclude this section, it can be said that representation has many symbolic and substantive benefits for minorities. It certainly is an indicator of political integration. The representation of minorities is often measured on the basis of proportionality to the population, a principle underlying employment equity, but it seldom reaches those levels. Representation is a necessary but not a sufficient condition for a political system's responsiveness to minorities' interests and needs. There are many other factors that also affect the realization of minorities' goals. Representation matters, but it operates within the confines of the political system and the long-established political culture.

Non-electoral Pathways of Political Incorporation

As this chapter is about the politics of inclusion, it is mostly focused on the representational aspects of political incorporation, more specifically on the electoral representation of ethno-racial groups. The other two elements of the policy framework described at the beginning of this chapter, namely, recognition and equity in services, are extensively discussed in chapters 4, 5, 9, and 11. Here I will examine non-electoral ways of participating in, influencing, and sharing in decision making that are part of the processes of political incorporation. Yet it must be pointed out that in the political arena, electoral and non-electoral modes of participation operate in tandem.

Community organizations, pressure groups, trade unions, corporate boards, chambers of commerce, media, business associations, political parties and clubs, faith-based institutions as well as advocacy groups are the non-electoral venues of articulating various interests, espousing ideologies, and mobilizing constituents to influence decision-makers.[41] Groups that can successfully mobilize these resources are significant players in the power game. Ethno-racial groups vary in their capacity, organization, and approach to marshalling these non-electoral resources. Generally, new immigrant communities have fewer such resources and are poorly incorporated. These factors change over time.

Except for relatively exclusive ethno-racial or religious organizations, most of the other civic associations bring together persons of different social backgrounds around common values and shared goals, for example, trade unions, business organizations, political clubs, and neighbourhood associations. They serve as the venues for political coalition building and bargaining across interests to build a common ground. In this sense the non-electoral channels of political incorporation also function as the instruments of mutual learning in the art of politics. They help immigrants in particular, and minorities in general, to learn the strategies of civic engagement and to expand their interests beyond ethnic politics to public policies.

Blacks in the United States have a long history of agitating for civil rights through organizations such as the National Association for the Advancement of Colored People (NAACP), the Congress for Racial

Equality (CORE), the Southern Christian Leadership Conference (SCLC), and the Urban League, which converged into the civil rights and Black Power movements of the 1960s. The history of this struggle is beyond the scope of our discussion, but it underlines the point that these non-governmental organizations brought about Blacks' liberation and gave them a foothold in the political structure.[42]

Jews followed a different, non-electoral path to their political incorporation. They built on their success in the professions, academia, the arts, and the media to play a prominent part in the politics of social causes (such as welfare and housing) in urban North America. Dominicans in New York have used the neighbourhood community council and the school board as the stepping stones to city politics. The point is that different groups carve different paths to amass political resources for participation in public affairs. In Los Angeles, business associations and labour unions have been the platforms for political action.

Canadian non-electoral pathways to political incorporation operate in the space provided by public institutions through their highly structured modes of civic engagement. Until the 2011 elections, visible minorities and immigrants relied largely on the Liberal Party for the advancement of their interests. The path of their incorporation went through the party branches at the constituency level. This changed in the federal election of 2011, when the ruling Conservative Party picked nominees in constituencies of ethnic concentration and helped them win. Toronto now has a robust network of community organizations and social agencies agitating for equity in services.

Globalization and transnationalism are opening new channels of political and economic influence in Canada and the United States. The Los Angeles area's trade and financial links with East Asia gave the Japanese community prestige and influence in the local economic power structure in the 1980s. In the 2000s, Chinese business and investment elites with connections in Taiwan and China have been on the rise. New York, Vancouver, and Toronto solicit Chinese investments in business and real estate.[43] The trade with Latin America has conferred on Latino immigrants the "middleman" role, which brings them some political influence.

Similarly, the opening of opportunities for dual citizenship has given ambitious immigrants, particularly businessmen, physicians, and high-technology professionals, the opportunity to circulate between their homelands and Canada and the United States. They parley their positions in one country into political influence in the other. Italian, Jewish, and Irish community leaders in North American cities have often drawn strength from their links in their respective countries abroad. Canadian prime ministers take prominent Chinese or Indian Canadians on their trade missions to China or India, for example. These connections abroad serve as political resources for immigrant communities.

The role of non-electoral modes of participating and influencing public policy has been long recognized in the political science literature. The notion of business and social elites forming the power structure for policymaking purposes has an enduring appeal. It draws its appeal from Marxist thought, though in political sociology C. Wright Mills gave a forceful expression to this concept.[44] Currently, urban regime theory picks up the argument for non-electoral contributors to local policymaking. Though it does not emphasize the notion of "power behind the power," it maintains that local policymaking involves coalitions of both elected leaders and influential persons in the non-governmental sector and media. Urban regimes are informal but stable coalitions that make key decisions for local communities.[45] Kristin Good, while finding regime theory a good fit for explaining the responsiveness of Canadian cities to immigrants' needs, also observes that "local leaders in both the public and private spheres create and implement local agendas."[46] The role of non-governmental leaders is explicitly recognized in regime theory.

The thrust of these arguments is that ethno-racial groups' political incorporation is a matter not only of elections to and representation in public institutions, but also of inclusion and influence in civic organizations, professions, and media. All in all, political incorporation is a process with many layers. It is possible for minorities to have a few elected representatives but a strong presence in the non-electoral segments of the local power structure and vice versa. This helps to explain the weak relationship in New York, Los Angeles, and Toronto between electoral representation and the steady accommodation of minorities' interests in city politics.

Summing Up the Politics of Inclusion

The question that underlies this chapter is how social diversity transforms the institutions of political power in multicultural cities. More specifically, how are the expanding ethno-racial minorities obtaining recognition, representation, and an equitable share of services and resources – the three components of political incorporation. This chapter suggests that the trajectory of minorities' political incorporation is upward in Toronto, New York, and Los Angeles.

The process of political incorporation is a matter of infusing diversity into political institutions and it follows the paths laid by the mainstream political system. Yet political incorporation does not proceed in a linear manner, advancing on all fronts simultaneously. It proceeds in spurts, advancing and reversing with economic and political cycles.

One indicator of ethno-racial groups' political influence is the responsiveness of cities' governance structures to their needs and concerns. Kristin Good has evaluated the responsiveness of multicultural urban and suburban municipalities in Canada to the minorities' interests. She found the cities of Toronto and Vancouver to be "responsive," the cities of Richmond and Surrey (Vancouver region) and the town of Markham (Toronto area) "somewhat responsive," whereas the cities of Mississauga and Brampton (Toronto area) were relatively "unresponsive."[47]

Susan Fainstein assesses West European and US cities on the criteria of equity, democracy, and diversity as indicators of a "just city." She found New York to be successful in accommodating diversity.[48] Similarly, Mike Davis's celebration of the Latinization of Los Angeles validates the city's inclusiveness.[49]

This medley of evidence shows that the accommodation of diversity has been woven into the current governance structures of multicultural cities in the United States and Canada. The politics of inclusion is working. Does it mean racism has disappeared or all policy biases have been eliminated? Of course not! The process of inclusion is always a work in progress. Yet it is an affirmation of multiculturalism.

Out of the three components of political incorporation, namely, recognition, representation, and fair share, recognition is largely a matter of discourse and symbolism. In multicultural cities, the discourse of diversity is pervasive. The city of Toronto has the motto "Diversity Our Strength," while New York is officially designated as "the world city of opportunity" and is described as "more cosmopolitan than Casablanca and [the] Shanghai of movies."[50] By proclaiming the national and religious holidays of many groups, by officially acknowledging group identities through proclamations, parades and festivals, by instituting anti-racism and employment diversity policies, and by attempting to be culturally responsive in the provision of city services, multicultural cities affirm ethno-racial groups' citizenship rights. These forms of recognition are almost ubiquitous in New York, Los Angeles, Toronto, Vancouver, Miami, and other such cities. Recognition has consequences. It legitimizes the equality of diverse groups and gives them justification for pressing their demands. It reaffirms identities.

Representation is essentially a political act, whether it is of the elected or appointed variety. It has to be won through elections, recruitment, or nomination. This chapter shows that the electoral representation of ethno-racial groups has increased over time, discriminatory barriers have been breached, and voters' power has found expression in elections. But as with women, ethno-racial groups seldom attain levels of representation proportional to their demographic weight. The political party structure, the divergence of interests and political ideologies, and the leadership and financial requirements of elections are intervening variables affecting the realization of electoral representation. Furthermore, proportionality itself is not a necessary condition for political incorporation. Non-electoral paths of representation are also sources of political influence. Ethno-racial groups differ in their capabilities to tap these sources. Some are more successful than others.

Amidst these trends, it is striking that New York and Los Angeles, the US cities, have higher levels of minority representation than Canada's proclaimed multicultural city, Toronto. The political system and culture of the two countries are different. In the United States, the civil rights and peace movements of the 1960s transformed the political culture and empowered grass-root initiatives. Ethno-racial groups are reaping the fruits of Blacks' struggle. The political entrepreneurship of ethno-racial groups is the primary instrument of obtaining representation in the US cities. In Canada's tradition of "peace, order, and good government," political institutions and public policies confer representation on new minorities, by and large. In the former, minorities' agency is the instrument and in the latter the structure endows representation.

As minorities come across barriers in the access to services, they demand accommodations and an equitable share. The representation and recognition are preconditions for the equitable distribution of services and resources.

Conclusion

Multicultural cities are politically inclusive. Ethno-racial groups' political incorporation follows the paths laid out by the political system and culture of the respective societies. Voters registering, joining mainstream parties, forming coalitions and alliances, mobilizing community power, cultivating political and professional leadership, launching media campaigns, and resorting to court challenges, protests, and agitations are among the tools of political incorporation. This is how multicultural politics is wedged into cities' political systems. Political representation, recognition, and equity in the distribution of resources and services are the outcomes of minorities' incorporation.

The role of the mainstream political systems as the common ground is to be underlined. Historical constitutions, public laws, and political values provide the structure within which ethno-racial groups seek their rightful place. Values, laws, norms, and practices change with new demands, yet the political system remains the medium through which divergent groups and cultures relate to each other. Through its construction and reconstruction, it persists and serves as the source of a cohesive political order for all. Multiculturalism does not undermine the functional imperative of such a structure. It helps new groups to be invested in political institutions. This chapter illustrates this proposition.

The relationship between political representation and responsiveness to minorities' needs and interests is weak. This chapter documents cases of moderate or low electoral representation and fair to high responsiveness to minority and immigrant needs. Yet representation gives a voice in the councils of power. Overall in the three cities, ethno-racial groups are underrepresented, some more than others. Race still matters, but over time progress has been made in gaining representation and recognition.

Canada and the United States differ in their political cultures. These differences are reflected in the modes of influencing the policies and patterns of minorities' political inclusion. In Canada, minorities are largely inducted into political institutions by political parties and through public policies. In the United States, minorities seek a fair share in power and then use it to bargain for responsiveness to their needs. Political incorporation is propelled by the drive for power and influence. In stark but over-generalized terms, the difference between the United States and Canada can be expressed as that between taking and receiving political influence.

There are notable differences among the politics of representation in the three cites, some attributable to the differences in national political cultures. Yet it is striking that New York and Los Angeles were further ahead in ethno-racial groups' representation in city government than Toronto and its suburbs. As multicultural cities, they have parallel institutions of incorporating diverse groups, but they differ in the modes and processes of incorporation.

The politics of multicultural cities evolves from ethnic to public-interest needs. Ethnic and racial communities discover that their self-interest is indivisible from the collective well-being. Alliances are formed on the basis of class, gender, lifestyle, or political beliefs over and above identity politics. This process of interest harmonization is led by mutual learning. And this is what takes place when ethnic representatives join city council or legislative assemblies. They find themselves pulled by the collective needs of a city as much as by the demands of their communities. They become aware of resource limitations and cognizant of social goals. Their transformation into public representatives sometimes is a disappointment for advocates from their communities. Yet political incorporation leads to the forging of a common ground.

Finally, the emergence of majority-minority cities is changing the political agenda. Increasingly, the issues will be of inter-ethnic competition for power and resources. The conventional narrative of minorities struggling against the majority's political domination is going to be less relevant. Instead, the political competition will be between one ethno-racial group and the others. How the competition among minorities is managed will determine the effectiveness of the political order in sharing power among, and meeting the needs of, different groups.

Chapter Nine

The Pluralism of Urban Services

The Cultural Coding of Urban Services

For living in a city, residents depend on a host of visible and invisible services, some provided by the public, others by the market, and still others by community organizations. City life is not possible without streets, drains, house numbers, water supply and waste disposal, public health measures, police, building and land use regulations, and so on. In addition, a contemporary North American city is inconceivable without electricity, telephones, TV, fire and safety services, schools, hospitals, community centres, parks, children and seniors institutions, environmental controls, social welfare, public transportation, housing for the poor, broadband and Internet facilities, and so forth. This web of hard infrastructure and soft human services stitches a city together and provides the stage for community life.

Whether a brick-and-mortar facility or a social program, every urban service is steeped in some cultural code, namely, rules, norms, symbols, and expectations by which a service or facility operates. This cultural code evolves over time, but it is deeply rooted in the history of the mainstream society and culture.

Even an apparently neutral facility such as the water supply operates on the basis of notions of the adequacy, safety, and cleanliness of water, behavioural expectations of an appropriate use of water, and contractual norms of keeping equipment in good order and ensuring utility workers' access to the metres and pipes on private properties. Cultural coding is all the more central to the organization of welfare, educational, children's, or health services. For example, the rule of eligibility for welfare payments has embedded in it the notion of the typical nuclear family of middle-class North America. It excludes dependent relatives living in the same household, as often found among the poor and immigrants. Similarly, the current definition of child abuse in the child welfare services is based on cultural norms that preclude any physical disciplining by parents, which as a cultural norm reflects the values of the liberal middle class. The point is that human services are visibly invested with cultural codes. Multicultural clients' needs and expectations are sometimes at variance with the cultural codes embedded in urban services.

The diversity of people has a bearing on both the demand and supply of services. On the demand side, ethno-racial differences are expressed in the divergent ways of meeting the same needs. On the supply side, first, it is a matter of equal entitlement to the available services. There are historic racial inequalities in the distribution of services whose legacy continues to affect minorities. Blacks have long been discriminated against in housing, schools, police protection, jobs, and health and welfare services, and have been continually lagging behind on these provisions.[1] Systematic disparity between the city services in low-income and affluent neighbourhoods often impacts immigrants and ethno-racial minorities in both Canada and the United States. The same was the situation for the early waves of immigrants in the middle twentieth century, for example, the Irish, Jews and Italians.[2]

Second, in the supply of services, ethno-racial differences have a bearing on the qualitative appropriateness of urban services. The cultural coding of a service may or may not be appropriate for satisfying the needs of different groups. Take the example of hospital food for patients: the usual North American cuisine – soup and a sandwich or meat and boiled vegetables – suits patients of northern European tastes. It would be unappetizing for Italians, Greeks, French, Chinese, Mexicans, or Indians. Also, Jewish patients and staff may require kosher, Muslims halal, and Hindus vegetarian food. The point is that the food policy of a hospital in a multicultural city will have to be both diverse and inclusive in taste to meet the needs of different backgrounds. In this case, equality of meeting needs requires differences in the types of food served. Cultural diversity has a way of bringing up such disparities in the demand and supply of many services.

New York's Parks and Recreation Department found that the food vendors licensed to sell snacks in parks and at sports events were not meeting the needs of park users in immigrant neighbourhoods. Local groups demanded such foods as tacos, tamales, and samosas and not just the traditional fare of hot dogs.[3] The department had to adjust its licensing policies. For the same reasons, Toronto has lately allowed the selling of ethnic foods from street carts.

These examples illustrate that uniformity of form does not satisfy diverse citizens' needs stemming from cultural differences. This is a situation of procedural equality leading to an inequality of outcomes. Here equality has to be realized by equalizing outcomes, rather than through a uniformity of inputs.

The discussion in this chapter is focused on the infusion of cultural diversity into the provision of urban services to make them pluralistic in form so as to equitably meet the needs of diverse populations. The questions addressed here include how the service systems of multicultural cities respond to the cultural and social diversity of their residents. What are the defining characteristics of urban services in multicultural cities? Again, I will be pointing out concepts and their indicators that will help guide empirical observations.

The focus is on observing the accommodation of culturally divergent modes of meeting public needs. This is not a comprehensive account of cultural adaptations in all the hard and soft services of Toronto, New York, and Los Angeles. Given the vast array of services and innumerable programs, it is almost impossible to cover all of these. Instead, the objective is to concentrate on patterns of accommodating cultural diversity in the provision of shared services.

It may be noted at the beginning that the process of accommodating ethno-racial groups' cultural differences unfolds incrementally as services are exposed to the culturally different ways of meeting users' needs. This is almost always a work in progress. None of the three cities has realized full responsiveness and equity in the provision of services. There are many other intervening factors (resources, technology, public-private balance in service production and distribution, etc.) that impede cultural responsiveness in the provision of urban services. I will focus on how cultural diversity is infused into the policies and programs of service provision.

Principles of the Production and Distribution of Urban Services

There are three principles to guide the organization of services in cities: availability, accessibility, and adequacy.[4] Each of these has a bearing on cultural differences in the provision of services in multicultural cities.

Availability refers to the quantitative provision of services for various groups. It is assessed in terms of the existence in sufficient quantity of a service or set of services to serve needs equitably. For example, the availability of housing for the poor is determined by the amount of affordable housing and the mechanism by which this is distributed. Affordable housing may be provided as public housing or via rent subsidies for eligible households. Are there enough public housing spaces or rent vouchers to match with size and type of eligible households? Such are the criteria by which availability is assessed. Availability is a prerequisite for accessibility and adequacy.

Accessibility is a matter of how potential users can access or be connected to services. It is a criterion to assess if a service reaches those needing it in terms of eligibility requirements, the language of delivery, affordability (if priced), and geographic location. This principle brings up the issue of institutional barriers that may shut out some groups or make it difficult for them to access services. For example, a policy of welfare payments may require that recipients be looking for work, which leaves out single mothers with infants or the disabled. Similarly, an employment assistance office located far from the residences of the unemployed may jeopardize their access due to the costliness of travel in both money and time. Regarding ethno-racial barriers to the accessibility of services, they could range from eligibility criteria to language barriers and discrimination.

Adequacy is a principle that overlaps the other two. It is largely a qualitative criterion that gives weight to the appropriateness of the forms in which services are delivered. It is based on the notion of satisfying the needs of clients in appropriate ways. Ethno-racial diversity requires that services be delivered in culturally appropriate forms to satisfy different groups. The example of hospital food is an illustration of the adequacy criterion. Adequacy implies tailoring the supply to the expectations of clients. It requires reasonable accommodations of different groups that do not impose excessive costs or compromise availability and access for others.

Racial and cultural biases in the delivery of services can take many forms. These biases may be overt in the form of racial and ethnic discrimination, of which there is a long history. And they can be systematic but unintentional biases arising from the managerial policies and practices of public agencies and private organizations. For example, by not recognizing foreign qualifications and experience, professional regulatory bodies end up barring immigrants from jobs in professions.

Similarly, the police profiling of minorities, mortgage bankers' redlining of Black areas, and poor schools in low-income neighbourhoods are examples of exclusionary barriers for non-Whites. Although US civil rights laws and the Canadian Charter of Rights prohibit discriminatory practices, which are not overt any more, eligibility criteria, market forces, patterns of geographic distribution, language, and modes of service delivery can combine to exclude minorities. Services have to be distributed in parallel with where and how people's needs are expressed and satisfied.

The Institutional Basis of the Supply of Services

There are three primary modes of delivering both hard and soft services, namely, via the market, public provision, and community organizing. The market as a supplier of urban services has a limited scope, as urban services do not conform fully to the model of price-mediated exchanges of goods and services. Many services are collective goods whose consumption is indivisible and often non-appropriable, making them difficult to transact in the market. How can police protection or street cleaning for some be separated from that for others, or traffic control be enforced through markets? Often using "the market" for many urban services primarily means the privatization of their production and distribution under contract from a public agency. And this is what happens when cut-backs hit hard services in particular.

Similarly, only a narrow range of soft services can be passed on to the market. Is there a market for marginalized youth programs, food banks, or human rights? Obviously, these services can only be provided through public provision. They may be contracted out to community agencies or private firms for delivery, but their planning and funding largely will have to come from the public. Urban services that can be priced, such as road maintenance, water supply, garbage disposal, and ambulance services, may be produced through the market mechanism, but given their externalities and welfare implications, they are generally organized as regulated utilities. Even they cannot be transformed into pure market activities.

Community-based efforts are the third mode of providing urban services. Pressed by the need for safety and welfare, human groups have always banded together to provide services on a self-help basis. In contemporary cities, communities of shared interests, geography, race, or culture come together to provide for child or senior care, youth recreation, neighbourhood watch, parks or street maintenance, and myriad other services. They constitute the third sector of the urban services system.

Community organizations also act as the implementing agencies for public programs. Cities contract out many services, particularly neighbourhood improvement programs or welfare and recreational activities, to community agencies. The City of Toronto, for example, spent $46 million in 2010 on community social services, out of which $17 million was distributed as grants to community groups for women's, youth, and marginal communities' programs.[5] In 2014, about $18 million was budgeted for this purpose.

When liberals or progressives come to power, the scope of services provided by the public sector widens; and as the political cycle swings towards conservatives, public provisions contract and the sphere of the market expands. Community initiatives are the default position between two ideologically charged modes of delivery. They can complement public provisions through grants and contracting out, but they also take up the slack from the market.

Cities in the United States and Canada have witnessed the cyclical expansion and contraction of public services over the last thirty years, both from political shifts at the national and local levels and from economic cycles of expansion and contraction. For example, the severe global recession of 2008–10 and the shaking of financial institutions in the United States have given a new incentive for the retrenchment of public services and shifting of responsibilities for the supply of services to the private and community sectors. These ideological cycles keep the availability and accessibility of services at the mercy of national and local politics as well as business cycles.

The modes of production and distribution of urban services have a bearing on responsiveness to the diversity of a population. Overall, public programs are relatively more amenable to the politically backed demands of ethno-cultural populations, not the least from the power of their vote. Many public services are based on the principle of universal provision, namely, coverage for the whole population with police, fire, ambulance, and public health services. Community-based provisions are relatively more responsive to cultural differences, because they are often initiated by communities themselves. Community organizations can be very successful on a small scale, but for bigger programs they depend on funding from the public sector or foundations. The market mode of distribution is usually less amenable to the demands of the poor and marginalized segments of a multicultural population. Yet it is flexible and responds quickly to cultural differences if there is sufficient demand from customers. For example, banks, insurance companies, private clinics, and lawyers readily offer services in multiple languages and employ co-ethnics if there are clients to be gained.

Cultural Responsiveness in the Provision of Services

Muslim women demand "women only" hours and female lifeguards for swimming in public pools. Hasidic-Jewish, Hindu, and Sikh women join them in such demands, and even feminists concur. Here is a case of cultural-religious values bearing on the adequacy of form in which recreational and sports services are delivered. In this case, the accommodation is easy, a matter of operational policies. Just set aside some hours for women. Yet it is never easy in cities of cultural differences. In the Montreal metropolitan region, some municipalities did set aside separate hours without any public controversy, while others sparked public debates about the "intrusion of religious values in the secular public realm and concerns of public hygiene," leading to the rejection of the requests.[6]

In multicultural cities, cultural responsiveness is an ongoing project. It unfolds as the historical forms of services are exposed to the demands arising from the cultural diversity of clients. First come the demands for accommodation on a case-by-case basis. Immigrants from the British Commonwealth countries begin to play cricket in parks, for example. They demand accommodations in the use of playfields. The second stage comes when accommodations build towards the harmonization of policies and revisions of rules. In the case of cricket, this stage comes with its recognition as a sport to be provided in the layout and management of playfields. It has happened in the Toronto area, where now there are twenty-seven cricket playfields and many more clubs. About twenty-eight schools have cricket facilities. New York City's Department of Parks and Recreation lists seventeen cricket playfields on its website. Los Angeles County lists fourteen such playfields. All these developments have occurred in the last twenty years. This is the phenomenon of the reconstruction of a common ground in recreation and sports.

Cultural responsiveness, a term referring to an institution's policies of accommodating cultural differences, is put into action through the concept of cultural competence. The two terms overlap to a large extent, though if one were to make a distinction, cultural responsiveness is the quality of an institution or organization, and competence is the skill or approach of a practitioner, which is also used by an organization. Cultural competence is defined as the practice of "delivering services that are congruent with behaviors and expectations normative for a given community and that are adopted to suit the specific needs of individuals and families from that community."[7] The US Developmental Disabilities Assistance and Bill of Rights Act Amendment of 1994 defines cultural competence as "services, support or other assistance that are conducted or provided ... in a manner that is responsive to the beliefs, interpersonal styles, attitudes, language and behaviors of individuals who are receiving services."[8] These definitions underline two ideas: one, the effectiveness of a service lies in matching it to clients' beliefs and behaviours including language, namely, through cultural competence; two, cultural responsiveness is a criterion for fulfilling the mission of many professions. It is a touchstone of multicultural practice in the provision of services.

Individuals responsible for planning and delivering services to diverse populations, particularly human (soft) services, are expected to acquire knowledge of the history, beliefs, and behaviours of their clients, be empathetic to their cultural differences, be aware of their own biases, and finally be able and willing to meet clients' needs in ways responsive to their cultures and languages.[9] These norms are now widely followed in multicultural counselling, social work, and psychiatry. In other professions, cultural competence is fostered through diversity training and mentoring by colleagues of ethno-cultural backgrounds similar to the clients'.

Culturally responsive agencies explicitly value diversity, organize their policies to meet the needs of a diverse population in accordance with ways suited to them, employ staff of diverse backgrounds, and have facilities to interpret clients' languages.[10] Of course, no single agency can serve the whole spectrum of cultures equally. Yet a culturally responsive agency recognizes differences even when not fully comprehending the norms, values, and language of a particular culture, and is able to draw on outside resources to deal with unfamiliar cases. Cultural competence may be the means, but the awareness of the need to accommodate cultural differences for effectiveness and equity is an end in itself.

At the organizational level, cultural responsiveness is enacted through the realigning of vision, values, and strategic and management policies and rules. A set of initiatives ranging from procedural to legislative, from awareness to investment (of resources), helps to enact cultural responsiveness in the provision of urban services. It is implemented in a series of steps to make service provisions pluralistic and inclusive, ranging from multilingual information and having a representative workforce to procedural accommodations leading up to institutional reorganization. These measures are arrayed in an ascending order of responsiveness in the following account.

Multilingual Access to Services through Translation and Interpretation Facilities

The portal through which services are accessed and used is language: that of information and communication. Obviously, persons who have difficulty understanding English or another official language cannot fully benefit from services offered by public or private agencies. Communicating with clients in their languages is the first step in the responsiveness of services. Translating documents, forms, and information bulletins as well as providing interpretation facilities to ethnic clients improves the accessibility of services.

Although there is public sensitivity about national languages, particularly in French-speaking Quebec and recently in California and Arizona, in practice non-English/French-speaking clients have often been accommodated through improvised interpretations. For example, a schoolteacher faced with non-English-speaking students calls in bilingual volunteers to assist. What multiculturalism does is to formalize the translation and interpretation services as a matter of right. It has happened explicitly in the governments of major Canadian and American cities.

The City of Toronto instituted the Multilingual Access Program in 1991, providing interpretation facilities in its public-serving agencies such as public health, the works department, ambulance, fire, and police services, parks, recreation, social services, and housing. The figures for the use of interpretation services are illustrative of the scope of these facilities; in the year 2000, oral interpretations were provided in fifty-two languages and written translations in thirty-seven. In 2002, the city expanded its interpretation and translation services by establishing a central Multilingual Service Unit and the Language Line Service. Interpretations of ethno-cultural languages are now provided in response to the needs of the target population or taking into account a neighbourhood's ethnic composition. These services are also routinely available in meetings and web-based consultations for public involvement in the city's decisions.

Yet not all municipalities in the Toronto metropolitan area fare as well in their multilingual policies. The City of Mississauga, at its feisty former mayor's insistence that immigrants should learn English, does not translate documents into other languages. The Town of Markham has interpretation and translation services. What this shows is that responsiveness to multicultural needs depends on local politics and community organization.

New York City has long supported the equal rights of minorities and a liberal outlook. It has always been a city of immigrants that take pride in its cosmopolitan culture, and revels in diversity. Yiddish, Italian, German, Chinese, and Spanish have long reverberated in New York's institutions and streets. A non-English speaker could almost always get services in New York on an ad hoc basis. In 2008, Mayor Bloomberg's executive order number 120 made this a right by legislating meaningful access to city services for all New Yorkers, including those with limited English. Each agency in the city is required to designate a language access coordinator, who has to develop an access policy and an implementation plan.

The Language Access Services must provide translation and interpretation, including telephonic interpretation, in the top six non-English languages, namely, Spanish, Chinese, Russian, Korean, Italian, and French Creole.[11] For other languages a roster of interpreters is maintained. A 2009 review of this policy showed that thirty-seven city agencies had detailed language access plans, including the Hospitals Corporation and Departments of Education, Information Technology, Children Services, Human Resources, and Mental Health. These efforts were estimated to cost about $26.9 million per year.[12]

Similarly, Los Angeles County has a Language and Culture Resource Center, which coordinates and responds to requests for interpretation and translation from county agencies in six languages, Spanish, Cambodian, Korean, Mandarin, Tagalog, and Vietnamese. It also provides sign language interpretation. Other cities in Canada and the United States also routinely offer interpretation facilities. The cities of Los Angeles, Chicago, and Vancouver and many of their suburban municipalities offer translation and interpretation services at public meetings.

The private sector responds massively to the opportunity for serving multilingual customers. Banks, insurance companies, real estate agencies, and merchants are quick to take up providing services in many languages.[13] Department stores post multilingual notices of sales and stock ethnic fashions and cosmetics. Hebrew, Chinese, Ethiopian, and many other languages beckon customers from store signs. Ethnic neighbourhoods have street signs and markers in their respective languages. Ethnic enclaves are the home of minority languages organized by community agencies. All in all, the private and community sectors respond opportunistically to the needs of multicultural communities.

From these accounts, it can be observed that, first, facilitating linguistic access to city services and decision processes is widely practised in multicultural cities of North America. Canadian cities institutionalize these practices more formally than do American cities, though New York is comparable to Toronto and Vancouver. Second, linguistic pluralism is grafted on English (French in the case of Quebec) as the official language, the latter retaining its role as the common medium.

Have these measures removed the linguistic barriers for minorities? Not completely, but they have increased their access and established the principle of accommodation.

The Representation and Involvement of Ethno-Racial Groups

Removing linguistic barriers is the first step on the ladder of cultural responsiveness. A policy of recruiting minorities, women, and persons with disabilities is the second step for enhancing the cultural responsiveness of an agency or institution. Staffing with persons of diverse backgrounds not only brings a better understanding of clients' needs, but also helps increase the satisfactoriness and effectiveness of programs.[14] The ongoing diversity training of the staff and policymakers further extends the scope of cultural competence in an organization.

An ethnic agency, a majority of whose staff is ethnically similar to clients, promotes ethnic identity, and includes ethnic content in its programs, takes the idea of cultural competence to its culmination. Afro-American-centred schools and Jewish or Muslims day-care centres are examples of ethnic agencies. Although there are some disquieting questions about the segregationist consequences of ethnic agencies, they are particularly useful in the delivery of services to new immigrants and marginalized communities.[15] By providing culturally and religiously appropriate services, ethnic agencies empower these groups and enhance the adequacy of services for them.

Yet it is the induction of minorities and excluded groups into decision making that consolidates the cultural responsiveness of service organizations. On this score, the past fifty years have witnessed steady progress in the enactment of anti-racism laws, human rights legislation, affirmative action policies, and employment equity programs. Minorities have not achieved full equality, but the barriers to their inclusion have come down.

The City of Toronto undertook a comprehensive exercise in instituting access and equity policies in its municipal operations in 1998. Not that, before this exercise, Toronto was a city of indifference to racial and ethnic inequities. The federal multiculturalism policy (1971) and Charter of Rights and Freedoms (1982) laid the basis for the introduction of equal employment programs. Toronto's 1998 policy essentially consolidated and institutionalized policies and practices that had been under way for a while.

Among the measures included in the Access and Equity Policy are the establishment of the Diversity Management and Community Engagement Unit, Employment Equity Office, and human rights monitors and, later, the Immigration and Settlement Policy Framework (2001). The sum total of these policies and offices is a concerted attempt to increase racial, ethnic, and gender minorities' representation in the city's governance and their involvement in decision making. Kristin Good, in a national study of immigrants' inclusion in city governance, observes about Toronto that "[its] political pluralism might be described as activist pluralism."[16] Toronto's activism is also the result of community agencies' strong role in advancing the involvement of minorities. There is a vigorous civil society advocating for immigrants' integration into the city's political and economic structure. For example, the Maytree Foundation has a program called "Diversity on Board" which connects qualified candidates of minority backgrounds with positions on the boards of public agencies, commissions, and nongovernmental organizations.

New York City also has employment-equity programs built on the long tradition of minorities' activism. It has an Office of Immigrant Affairs to provide for immigrants' service needs, which coordinates with city agencies to facilitate their accessibility. It provides training to immigrant businesses and leaders of immigrant communities. Minorities' representation in the city's civil service and political leadership has been increasing. The increasing representation of ethno-racial groups on the city council has been described in chapter 8. Their representation in administrative positions has also increased. In 2014, newly elected Mayor de Blasio's executive appointments increased the share of minorities to 55% and that of women to 54%. All in all, the representation of minorities in decision-making positions is a work in progress. It may not have eliminated inequalities and the sting of discrimination may still be felt, but the legislative ground and political commitment for bridging those chasms have been laid. Still, the involvement of diverse groups in decision making is subject to the political winds blowing at a particular time.

The City of Los Angeles elected Thomas Bradley as its first black mayor in 1973. He held office until 1993. In 2005, the city elected its first Hispanic mayor since 1872, Antonio Villaraigosa. They symbolize the opening of the power structure to ethno-racial minorities. With the city and region demographically becoming areas of majority-minorities, they have established employment equity policies, anti-discrimination legislation, and many programs of citizen participation and the inclusion of diverse communities in decision making.

The history of affirmative action in California is full of reverses, as exemplified by Proposition 209, voted on in a state-wide referendum banning preferences given to minorities to make up racial and ethnic shortfalls in representation, and upheld by the state's high court in 2010. Still, the notion of equal opportunities for ethno-racial minorities in employment and services continues to rule.

In rounding up this section's discussion, it may be pointed out that we have not evaluated the outcomes of such widely spread out measures. The general observations culled from the foregoing discussion are (1) staffing with persons of backgrounds similar to those of the clients is another measure to realize the cultural responsiveness of services, and (2) the multicultural cities of Toronto, New York, and Los Angeles have introduced policies and established agencies to promote employment equity and the representation of minorities; their measures are reflective of countrywide trends.

Accommodations of Ethno-racial Differences

When a school gives extra time to someone identified as a "special needs" student to complete a test, it is an accommodation. As a deviation from a convention or rule which adversely affects or disadvantages a person or group, an accommodation is a much-used practice to adjust for differences of ability, gender, race, or culture.

From the ethno-racial perspective, there are two types of accommodations in service provision: (1) making allowances for differences in
the existing forms of service; (2) offering special services for groups
with specific needs, such as immigrants, racial-minority youth, refugees, or disabled persons. There are countless examples of both types.

Ethno-racial differences are often accommodated through changes in
the content of services or the provision of special services in a variety
of situations. Examples include myriad practices: setting aside women-
only swimming hours in public pools, allowing observant Muslim girls
to play soccer with head scarves, amending city by-laws to allow non-
English street signs, enacting anti-racism policies for the police force,
subsidizing ethnic-specific seniors' homes, day-care centres, and housing cooperatives, hiring co-ethnic counsellors for natives, permitting
saris as the office attire for South Asian women, and providing prayer
rooms in offices.[17]

An interesting example of helping to develop cultural competence
among New York's police personnel is a manual prepared by the police
academy entitled "Policing a Multicultural Society." It points out the
manners, speaking habits, gestures, and other cultural behaviours of
different groups to sensitize police officers in dealing with people of
diverse habits and temperaments.[18]

Among the special services available for ethno-racial groups are language classes, job-search assistance centres for immigrants, and legal
aid for housing. A striking example of ethno-cultural accommodation
is found in library services. Most North American cities' local libraries
now offer books, newspapers, videos, and computer resources in the
languages of their resident populations.

First Nations people, racial minorities, and ethnic communities have
special needs arising from their cultural and religious dispositions or
barriers in their access to existing services. Special programs are continually started, revised, and abandoned as part of the business and
political cycles. Successive cycles gives rise to experiments such as
minority-youth-at-risk counselling and engagement services, programs
for single mothers' education, Afro-centred schools, ethnic festivals,
theatres, folkloramas, and parades to highlight the heritage of minorities. The point is that attempts to accommodate the special needs of
ethno-racial minorities have become a part of the service strategies of
multicultural cities in Canada and the United States. How adequate are
these and do they help bridge inequalities of access? Undoubtedly, the
needs outstrip the supply and inequities persist. The imbalance is also

sustained by the continual inflow of immigrants who particularly need accommodation in the provision of services.

All in all, accommodations are wide-ranging, numerous, and the most frequent way of responding to differences in needs. Cumulatively they transform the provision of services from a unitary to a pluralistic form. This is how diversity is woven into service institutions of multicultural cities.

The Harmonization of Practices and Reconstruction of the Common Ground

The sum total of accommodations is the restructuring of policies and practices, making them pluralistic, accessible, and adequate for both the mainstream and minority communities. It leads to the process of harmonization, which is a step beyond individual accommodations. Harmonization is the process of revising and adjusting norms, rules, practices, or policies to reconcile the expectations of different communities. Gerard Bouchard and Charles Taylor call it a form of concerted adjustment of norms, values, and practices.[19] It reconstructs the common ground to make it more inclusive. The aim of a revised policy is to be equally adequate for all. Some examples will illustrate the processes of harmonization and reconstruction of the common ground.

Educational institutions in North America have long faced the demands of recognizing the histories and cultures of racial and ethnic minorities so as to provide equal opportunities for such students and staff and accommodate their sociocultural needs. The response to these demands has been on many fronts, from changes in curricula, teaching methods, school management, and community involvement to innumerable accommodations of cultural and racial differences. Local school boards are the arenas where these issues play out. There are thousands of measures under way at any one moment in North American cities.

The accommodations could be as generic as providing breakfast to all students in poor neighbourhoods and offering courses on Black, native, or ethnic cultures, or as specific as providing language and cultural interpreters for immigrant youth, allowing prayers rooms in schools, holidays for Jewish, Muslim, Hindu, and Sikh students on their holy days, accepting wearing of the hijab in classes and sports, and many similar adjustments in practices too varied and numerous to recount. Many school boards, colleges, and universities pull these strands of accommodation into generalized system-wide diversity policies addressed to different situations. This harmonization of episodic accommodation results in establishing school-district- or university-wide visions, norms, and values which are the basis of a revised common ground.

The harmonization of policies to accommodate diverse needs is reflected in the recent executive order of Mayor de Blasio of New York to give holidays in public schools of the city on the two Eids of Muslims. The schools will be closed on these two days, though to maintain the total number of study days, summer vacations will be correspondingly reduced.[20]

The Toronto District School Board (TDSB) has developed a vision statement called "Equity Foundation," which enunciates norms and values of "equity of opportunity and access," while recognizing "systematic biases related to race, colour, culture, ethnicity, class, age, ancestry, religion, gender, sexual orientation, marital status and thus commits itself a) to ensure that *all* students understand the factors that cause inequity, b) *all* learners are provided support and rewards to develop their abilities, c) promote *equitable representation of diversity at all levels of the school system, etc.*"[21] (italics added). The point of citing this policy statement is to illustrate the inclusive common norms and values that emerge with the harmonization of educational practices.

Health services, particularly mental health services, are another area where a harmonization of policies and practices is at the forefront. A new approach to healthcare has emerged, based on the social determinants of health, to promote responsiveness to the social and cultural elements in health services. Awareness of the higher susceptibility to hypertension, cancer, and diabetes of Blacks and Asians and of culturally mediated symptoms of anxiety, depression, and schizophrenia, for example, are influencing healthcare practices and public health programs in cities of diverse populations. These ideas are introducing new norms for health providers' behaviours and the interaction between clinicians and patients. These norms take account of diversity while laying the bases of a pluralistic approach to health services.

The kirpan, a small knife or dagger ceremonially worn by male Sikhs as a religious requirement, is an illustrative case of harmonization in school policies in California and Quebec. This harmonization has come through judicial adjudication after a series of accommodations. The bare-bone facts of the cases are that Sikh children demanding to wear the kirpan as a matter of their faith were accommodated in the same manner by two school districts, one in Merced County, California, and the other in Montreal, Quebec. Both districts, though in different countries, imposed restrictions on the kind of kirpan and how it is to be worn, namely, it be a dull blade only a few inches in length and sewn into a sheath. Yet the religious requirements of Sikhs were accommodated, despite policies of banning knives on school premises. The schools' accommodations were overturned by the school board or local court. Both cases ended up before appeal courts. Quebec's case went up to Canada's Supreme Court, and was argued on the basis of the Charter of Rights and Freedoms. The case in California was argued on the basis of the US Religious Freedom Restoration Act (1993).[22] In both cases Sikh children's right to wear the kirpan as a religious symbol, under strict conditions, was upheld, establishing a new norm of religious tolerance balanced with rules for safety.

The main conclusions suggested by this section are as follows: (1) the culmination point of cultural responsiveness is pluralism of policies and practices; (2) which results in a redefinition of norms and values aiming to realize the equity of outcomes for people of diverse backgrounds; (3) in turn leading to making institutions inclusive, within the parameters of their missions and functions. (4) An incremental process of reconstruction of the common ground unfolds, while its fundamental values are sustained and reaffirmed.

The Politics of Pluralistic Service Provision

What do the measures described above add up to? They represent a growing cultural responsiveness in the discourse and ethos of urban services. It has led to many changes in the provision of services, but the persistence of disparities for ethno-racial minorities suggests that the discourse does not entirely translate into action.[23] Yet the discourse has laid the moral and legal ground for minorities to claim accommodations and demand the restructuring of institutions. Equality of outcomes will remain a process rather than a condition. The gap between the supply of and demand for services suitable for minorities will continue to exist. There are many reasons for it, which lie in the broader realm of evolving political, economic, and technological conditions.

Economic cycles of expansion and contraction sweep across urban services, increasing their supply during one phase and shrinking their availability in another. Also, ideological swings between conservative and liberal political regimes produce periodic expansions, privatization, and retrenchment of public services.

Recently, the fiscal crisis of 2008–10 has severely set back public programs and services in both the United States and Canada. Budget deficits and a shrinkage of revenue not only squeezed the availability of public services, but also curtailed the availability of services provided by community organizations and the private sector. In 2011 about forty-six states in the United States were plagued by budget deficits; their shortfalls were projected to grow to $113 billion by 2012.

Canada's three largest provinces, Ontario, British Columbia, and Quebec, were also running deficits of billions of dollars. Almost every large city in North America was in financial stringency; some even toyed with the idea of declaring bankruptcy. In these circumstances, concerns of equity and cultural competence fall off the political agenda. Obviously, cultural responsiveness and ethno-racial equity cannot be pursued single-mindedly, without any regard to the overall state of services and budgets. Yet all is not lost. The notions of responsiveness and equity are now deeply entrenched in the professional ethics of service providers.

Apart from the political economy of the service systems, there are internal technological and organizational factors whose change affects the production and distribution of services, for instance, the Internet-based delivery and administration of some services.

My argument is that responsiveness to diversity and equity is one among many objectives. It is constrained by other considerations. Therefore, an imbalance between the ideal and the reality of cultural responsiveness is an enduring condition. This does not imply that the discourse about equity and responsiveness in the provision of services is fruitless.

The discourse about cultural responsiveness and harmonization itself is not unchallenged. The political right keeps bringing up arguments about national fragmentation and reverse discrimination to challenge policies of accommodating minorities. They advocate for colour- and culture-blind access for all, while overlooking the built-in structural biases of service systems. The charge of preferential treatment of minorities flies around and feeds a political backlash. Recently, the tide has been turning against affirmative-action policies. The US Supreme Court in 2014 restricted affirmative action in admissions to schools.[24] In 2009 the Supreme Court invalidated New Haven, Connecticut's preferential promotion of non-White firemen over twenty White applicants.[25] Canadian human rights tribunals are inundated with claims and counterclaims of discrimination by persons of both the minority and majority backgrounds. The politics of fairness both promotes and constrains cultural responsiveness in the provision of services.

Finally, the geography of a city reflects spatial disparities in the provision of services. Poor neighbourhoods and the minority residential areas tend to be ill served in housing, schools, police, public health, and transportation, and so on. Racial ghettos are a standing testimony to the inequities of service provision. These disparities require sustained public investments and culturally responsive policies, which are coming about contingently and incrementally.

Mainstream Institutions as the Common Ground

Accommodations, cultural responsiveness, and other measures are meant to tailor the delivery of services in ways that satisfy people of diverse backgrounds. Yet they are rooted in the existing institutions, policies, and practices that form the service infrastructure of the mainstream society, which serve as the common ground or base. Mainstream policies and regulations are active agents in the process of infusing diversity into service provision, not barriers in the realization of equity for diverse groups. Not only do they incorporate new values, norms, and needs, but also they define the scope and limits of the infused elements. They are meant to advance the common purposes and shared interests for which the respective services are provided. They inject common objectives into ethno-racial demands and fulfil the shared goals of public services. They accommodate as well as set limits to the accommodations made necessary by ethno-racial differences.

For example, ethnic restaurants and food carts in following customary ways of preparing and displaying food have been given many accommodations in the three cities, but public health regulations about cleanliness, cooking, and storing practices are enforced to prevent food contamination for the health of consumers. In Toronto, New York, and Los Angeles, ethnic eateries are inspected, approved, and rated on the basis of their cleanliness and food preparation methods. These regulations promote food safety and inject universal norms and scientific knowledge into ethnic food businesses.

New York accommodates ethnic and religious festivals and holidays by suspending parking restrictions on such occasions for everybody, but it maintains the integrity of the parking regulations for the common good of traffic management.

These examples show that mainstream institutions are founded on pluralistic modes of delivery. They are transformed in this process, but retain their structural framework to maintain coherence and compatibility among different modes of delivery. They set limits to cultural relativity and establish a common base of services for all citizens. Mainstream institutions are the medium for micro-negotiations of differences and mutual adaptation across cultural differences. They sustain the coherence of urban services and reconcile differences, themselves changing in the process. Yet such outcomes are not always realized. They are highly dependent on local politics and resources.

Conclusion

Pluralism of policies and practices in the provision of urban services is part of the transformation process that creates a multicultural city. It involves the infusion of cultural responsiveness and equity of outcomes into the institutions delivering urban services. These policies are expressed through practices such as interpretation and translation into minority languages, accommodations in policies and programs, representation of co-ethnics among service providers, and the harmonization of systems of service delivery. They come about by political and administrative actions in response to ethno-racial groups' demands.

This chapter has examined the patterns and processes of infusing a multicultural ethos into the provision of services in multicultural cities. It has not given a comprehensive account of the accommodations effected, nor does it find that all demands are equally realizable. The accommodations and harmonization are occurring in piecemeal and reactive ways. They are not conceived as a broad strategy for a city's whole infrastructure and services. From the cases cited, it appears that social, health, and educational services are more readily adopting cultural responsiveness. By contrast, hard services such as fire and those having a strong component of universal-legislative coverage (housing codes, traffic laws, eligibility for unemployment allowance, etc.) are less amenable to accommodating cultural differences. Overall, the discourse of cultural responsiveness is more advanced than the practice. There are structural conditions, technological challenges, and historical inequities that impede the realization of the ideals of pluralism and equity in the actual provision of services.

Mainstream institutions, organizations, and public values are the foundations of the urban services on which accommodations are raised.

They have a critical role in defining the scope of and limits to infusing cultural responsiveness. They are transformed in the process of injecting a pluralistic ethos and in turn they instil common values in ethno-racial groups' demands.

Pluralism of services fits into a larger pattern of integrating diverse groups, including ethno-racial minorities, women, and the disabled, into a city's social, economic, and political institutions. The social sustainability of a city lies in adopting policies and reconfiguring governance structures to be more inclusive of diverse groups. Municipal governments play a critical role in this regard. Through pronouncements, strategic policies, and regulations, municipal governments lead the way in developing a multicultural ethos in the public as well as private sectors.

Chapter Ten

Urban Planning for Cultural Diversity

The Basis of Multicultural Planning

Urban planning is a professional activity meant to guide and manage the development of cities. It is a primary instrument of policymaking and management in cities. A great variety of urban issues, be those of land use, physical design, housing, community services, economic development, environment, transportation, and infrastructure, are addressed through urban planning. The solutions to these issues are meant to be based on public values such as efficiency, equity, safety, health, welfare, sustainability, and beauty.[1] Ethno-racial diversity injects elements of cultural difference and social differentiation into the conception of problems and the interpretation of these values. This is the ground where multicultural planning grows.

How are cultural differences and racial equality incorporated in planning policies and programs? How is the planning system used to reconcile cultural differences with the public interest?[2] This chapter probes these questions by examining the planning responses to ethno-racial diversity in North American cities, but particularly in Toronto, New York, and Los Angeles. It examines what has been done and can be done to incorporate diversity in planning policies and practices.

A large body of literature, including the feminist corpus, has emerged pointing out the limitations of policies and programs based on a presumed universalism of norms and values that essentially project the needs of dominant groups.[3] In the same vein, the idea of public interest is recast to "reflect the diversity of interests" out of which the common good may be "established discursively," with the participation of multiple publics.[4] Social and cultural diversity requires accommodating differences of urban needs and weaving, out of a variety of interests, plans that promote shared goals of efficiency, equity, sustainability, and convenience for urban development. The two-edged challenge of planning multicultural cities includes recognizing and accommodating differences, but also forging and strengthening a liveable city shared by all. A few examples illustrate the scope of this challenge.

Competing Interests and Divergent Needs

Almost every day, somewhere in Toronto, Vancouver, New York, or Chicago, there is a community meeting about building a suburban mall, establishing a mosque, or organizing a shelter for the homeless. In these meetings, divergent conceptions of what is needed and how to develop it emerge, reflecting the differences of class, culture, and race of the proponents. This was what bedevilled the urban renewal programs of the 1960s and 1970s,[5] and it continues to resonate in current controversies concerning gentrification and area development. With immigration and the consequent explosion of ethno-racial diversity, the differences of community interests have also taken on the cultural colours and politics of identity.

An Asian mall is proposed in the San Gabriel Valley, Los Angeles County, and political battle lines are drawn between the Chinese community and the long-established White suburbanites. An eruv, a symbolic enclosure of an area within which movements are permitted on the Sabbath, is erected by orthodox Jews by stringing wires on poles in a neighbourhood. In New York and Toronto, eruvs caused no controversy. In Montreal and Tenafly, New Jersey, they led to long, drawn out litigation.[6] Socio-cultural differences surface even in such matters as the definition of family in housing and zoning regulations. The historic inequities of Blacks' needs and the neglect of poor neighbourhoods also call for urban planning that fulfils its high-minded goals.

The point is that socio-cultural differences give rise to divergent demands, which necessitate a pluralistic rather than unitary (one-model-fits-all) planning approach to urban development. Also, community interests, defined by a combination of race, ethnicity, class, and gender, compete for resources and space. They call into question prevailing conceptions of problems and solutions. Yet the multiplicity of demands for space and services ultimately has to be reconciled into an integrated approach for the common good. These points highlight how diversity operates within urban planning.

Equality Frames Diversity

It is a truism to say that multicultural planning rests on two values, equality and diversity. The scope and meanings of both these values have been discussed earlier in this book. At present I want to reiterate the distinction between equality and equity. Equality is the right or entitlement to equal status under the law and equal benefits of access and treatment, without discrimination. Equity is the outcome of incorporating equality rights in an institution, policy, or activity. The right of equal employment opportunity is concretized as equity in recruitment and promotions. Equity may require modifying "the consequences of a strict application of the law to avoid unfair or unconscionable outcomes."[7] Equity is also a form of enacting fairness in the outcomes of distributing opportunities and resources. It has an affirmative and accommodative thrust. Accommodations advance the objectives of equity.

One distinctive feature of equity in urban planning is that it extends to group interests above and beyond those of individuals. In matters of urban space and services, many facilities are group-based and an individual cannot have access to them unless they are available to a community or group as a whole. For example, equitable access for a Muslim to a place of worship means there has to be a mosque for a congregation. So equity in this case is realized if mosques are included in city plans and zoning by-laws in inclusive regulations of places of worship. A similar logic applies to cricket players, bocce bowlers, or physically handicapped persons claiming equitable spaces in city parks. The equity for an individual is only realizable if a group is treated equitably. Thus, cultural differences in urban planning largely are expressed in group facilities and services. This observation brings up the question of how ethno-racial equity is incorporated in urban planning programs and policies. This question calls for examining both the theories and practice of planning.

Multicultural Practice and Planning Theory

There are two distinct views of how urban planning responds to the demands of diversity and equality. Planning theory, by and large, views urban planning as relatively unresponsive to people's rights to differences, and the recognition of their cultural needs thereby overlooks their entitlement to equality.[8] Planning practice points, as proof of its responsiveness, to the vibrant multiculturalism of North American cities, a thriving ethno-racial diversity in both public and private spheres, and accommodations of socio-cultural differences in policies and programs.

In the planning literature, the topic of planning for diversity or multi-cultural planning is discussed essentially in the form of critiques of and exhortations for urban planners, past and present. Leonie Sandercock draws attention to the "glaring absences in the mainstream accounts of planning history," namely, the voices of ethno-racial minorities, women, and those lacking in power.[9] Dory Reeves advises professionals to "value diversity, promote equality and become more conscious of power relations."[10] Most of the writings on multicultural planning are couched in prescriptive terms, describing what should be done rather than what is the practice.[11]

The following five propositions summarize the current theoretical discourse on multicultural planning. They encompass almost the full range of planning theorists' arguments about urban planning's responsiveness to socio-cultural diversity.

1. Urban planning is coded in Anglo-European cultural precepts. Those are held to be the universal norms of people's needs and preferences. The built environment is inscribed with these so-called universal precepts, privileging the culture of the dominant community. Urban planning is the agency for doing so.[12]

2. The modernist bias of planning, its "enlightenment epistemology," rational-positivist approach, and scientific analytics have been held to be a barrier keeping out the voices and stories of minorities, women, and other citizens.[13] Leonie Sandercock calls the current approach a "heroic model of modernist planning" in which rationality, comprehensiveness, the scientific method, and faith in planners' ability to know what is good for people generally and political neutrality come together.[14]

3. Another theme that courses through academic writings concerns planners' and their political masters' lack of sensitivity to cultural rights.[15] It is expressed either directly as a drawback of the planning system, or indirectly as exhortations for planners to be sensitive to racial-cultural differences. Its soaring rhetoric places planning practice in the context of colonial attitudes towards aboriginals, racial minorities, and immigrants.[16] Thomas Huw, in more measured terms, locates the understanding of planning practice in the histories of Western countries and their legacy of racism.[17]

4. The theoretical discourse is rich in proposals about reorienting planning practice and turning planners into activists and advocates for the rights of minorities. The emphasis is on the values of planners and organizational culture, calling for "new modes of thought and new practices" and shifting away from "outmoded assumptions embedded in the culture of Western planning."[18] Planners are advised to be politically committed, personally audacious, professionally creative, and socially therapeutic.[19] In radical visions of planning practice, planners are challenged to be insurgents in organizations and the vanguard of social change for a just city.[20] These formulations recapitulate planning theory's discourse of the heady days of advocacy planning in the 1970s.

5. Finally, theorists focus almost exclusively on the procedural aspects of urban planning. Their criticism and ideas for reform are conceived in terms of processes of planning decision making.[21] They envision planners as the agents of empowering marginalized groups and as promoters of their involvement in making plans.

Obviously, theorists do not credit planning practice as being responsive to cultural and racial differences in needs. But planning professionals view their work to be responsive to the demands of multiculturalism within the limits of their mandate. They do not write much about their ideas, but point to the multicultural vibrancy of cities as evidence of their responsiveness. Domenic Vitiello observes that "planners and community development practitioners have incrementally addressed many problems immigrants face."[22] It might be also noted that in the roster of immigrants' and minorities' discontents with "the system," planning policies seldom appear, except in terms of neighbourhood issues or disagreement about the use of a site. What has been happening in practice in multicultural cities to incorporate diversity is the test of the responsiveness of urban planning.

The Planning System and Multiculturalism

Contemporary urban planning, particularly in the United States, has been influenced by the long struggle of American Blacks for fair housing, community control, equitable services, and citizen participation in decision making. Paul Davidoff's seminal article "Advocacy and Pluralism in Planning" transformed the conception of urban planning.[23] Appearing in the midst of the civil rights movement, it presented a planning model that acknowledged the diversity of community interests. It put forth two ideas: (1) multiple (pluralistic) plans, each prepared from the perspective of a particular group or interest, and (2) public advocacy of plural plans and negotiations among competing interests on the bases of their plans to forge a balanced proposal. Advocacy planning spawned citizen participation as an integral part of the process of planning. It has, over time, opened the process of plan making to participation by lay communities in all stages of planning and implementation.

Today's planning is a collaborative and communicative exercise, in which citizens, including minorities, women, and immigrants, who are stakeholders are meant to be involved in preparing and implementing plans.[24] By and large, these practices are not much different from what theoreticians of multicultural planning propose. Planning theoreticians, in writing about multicultural planning, tend to concentrate on the processes of plan making, and not much on the actual policies that form the substance of plans. This is a well-recognized bias in planning theory which Michael Brooks views as "the process component of our profession."[25] Richard Klosterman comes to a similar conclusion about the limited focus of planning theory, based on a survey of the theory courses in US planning-studies programs.[26]

The challenges planning practice faces are of balancing the demands of multiculturalism with the goals of health, safety, prosperity, sustainability, and other collective goods necessary for an area. It also has to reconcile competing ethno-racial interests vying for the same sites and services. Its pursuit of equity has to contend with both social-class and ethno-racial inequalities, which sometimes may diverge from each other. The point is that urban planning is not a one-dimensional activity. It has to cater to multiple objectives within the scope of its defined institutional mandate.

As an institutionalized activity, urban planning has a legislated mandate that varies between Canada and the United States, and within these countries, among provinces and states. Yet there is a common core of objectives and instruments. In both countries, urban planning largely deals with land use, transportation and infrastructure development, policies for housing supply and management, programs for community services and economic development, environmental sustainability, and the provision of recreational and cultural facilities.[27] It is local and regional in scope and has a limited range of instruments to influence public policies and markets. It is an institutionalized activity, with defined powers and functions. It does not range over all societal and state activities. In theoretical discourses about multicultural planning, these institutional limitations of urban planning are ignored. In fact, the use of the adjective "urban" is often avoided and planning is presented as a generic function of societal management.

Professional planners draw their authority from local, regional, and provincial/state legislation and they work under the direction of elected councils, executives, and autonomous commissions or committees. Their recommendations are often modified and overruled by politicians. Citizens' voices and inputs have significant influence over planning decisions, but no one, whether citizens, politicians, administrators, or professionals, has veto powers. Planning processes work through the inputs of various actors to arrive at decisions. Planners are central, but not the only actors in decision making. In the American planning system, mayors and councillors have more say than in the Canadian system, which revolves around legal-bureaucratic institutions. Property rights are stronger and the market plays a greater role in the United States. The responsiveness of urban planning to diversity has to be assessed within the scope of this institutional framework. One cannot assess urban planning on the basis of objectives that are not part of its institutional mandate, such as being an "insurgent planner."

My focus here is on the institutionalized but comprehensive practice of area-wide planning. Yet there is community planning outside the institutionalized activity of urban planning. It occurs in neighbourhood associations, community agencies, and other non-governmental organizations for particular functions (such as neighbourhood improvement, business development, or protecting tenants' rights) and the advocacy of community goals. I will refer to this level of planning practice later. For now, the focus of discussion is on how urban planning accommodates diversity.

The Inclusiveness of Planning Processes

One indicator of the responsiveness of urban planning is the participation of socio-ethnic groups in planning decision making. It takes two forms. The first invites, solicits, and facilitates expressions of the interests and concerns of socio-ethnic groups in policy making and implementation. The second empowers members of minority communities and brings them in as staff, managers, and elected/appointed public representatives. The first level has been largely realized in urban planning.[28] Regarding the second level, the inclusion of socio-ethnic groups has advanced, but not fully. Chapter 8 showed that, politically, some groups have gained adequate representation while others lag behind. Generally, racial barriers have been breached, but racism has not disappeared. With multiculturalism, new issues of inter-ethnic competition are arising.[29]

A survey of forty-two cities, twenty-three in the United States and nineteen in Canada, based on the policy index (to be discussed later), shows that providing translation and interpretation services in ethnic languages, involving ethno-racial organizations in planning deliberations, appointing minority representatives to task forces and planning committees, and diversifying planning staff are among the most frequent practices followed, particularly in large cities of 500,000 or more.[30] What this survey captures is urban planning's inclusion of citizens' interests and concerns in the formulation of policies and plans.

Urban planning has been at the forefront of democratizing such decision making. Its responsiveness to minorities has been woven into its practices since the days of advocacy planning. Minorities' participation is facilitated through accommodation of their linguistic needs and representation of their voices. Innovative measures are increasingly being used to involve stakeholders, such as meeting in small groups in people's homes, using Facebook, Twitter, and other forms of social media, and holding design charrettes.

The City of Los Angeles, for example, held sixty community and neighbourhood meetings and involved three thousand residents and business people during the preparation of its general plan from 2000–2. The city provided interpreters, translators, and planning documents in Spanish, Japanese, Korean, Chinese, and Persian for the respective groups. New York's mayor issued an executive order for all city agencies to provide language help for non-English speakers, while minority community organizations are given grants for their activities. In addition to interpretation services, the City of Toronto funds community organizations engaged in anti-racism initiatives and assists ethno-racial groups in contributing to the city's policy discussions.[31] These cities and others also have diversity and employment-equity programs for recruiting staff. Vancouver has created a special position of multicultural planner to advocate for diversity interests. Toronto city's Social Development Division carries out audits of the city's agencies to assess their conformity to its access and equity guidelines.

The point of these examples is to show that the institutional basis for inclusion of the interests and concerns of ethno-racial groups is well developed in urban planning. The outcomes of these measures may not satisfy all groups, but policy making involves political bargaining and trade-offs among competing interests. Politics and power influence planning decisions. Those minorities that have mobilized their voting and political power, have gained influence in policymaking. All in all, ethno-racial groups are not entirely shut out of planning decision making, but have varying degrees of influence, depending on local conditions and community mobilization.

Negotiating Cultural Differences over Land Uses and Community Institutions

In an urban landscape, socio-cultural differences are expressed in land uses, residential and commercial developments, public institutions and architecture, as well as signs and symbols. To gain an understanding of how cultural differences are articulated and accommodated in the urban landscape, a review of some typical multicultural developments will be illustrative, beginning with the examples of non-Christian places of worship, namely, mosques, temples, or gurudawaras, in North American cities. The demand for these facilities is relatively recent, originating with immigration from Asia and Africa.

The development of mosques has become a matter of some controversy, particularly after the September 11, 2001 attacks on the World Trade Center in New York and on the Pentagon in Washington. A wave of anti-Islam sentiments has surfaced, which has led to sporadic citizen resistance to the development of mosques. In this political environment, the development of mosques is a good test of a planning system's responsiveness in dealing with minorities. On this test, urban planning has fared well. It has tried to isolate development issues and address them within the scope of its mandate.

A case in point is the proposal to develop an Islamic community centre and mosque in an existing building, 51 Park Place, near Ground Zero of the destroyed World Trade Center. New York's Lower Manhattan Community Planning Board recommended the proposal in May 2010, by a 29 to 1 vote with ten abstentions. After long and acrimonious public discussions and protests it maintained that the proposals had met all the planning criteria and held the religious-political objections to be extraneous to planning decisions. This decision sparked countrywide protests and media campaigns and became an issue in the 2010 congressional elections. Yet the city planning department held firm, with the support of the mayor. An attempt to block the approval by getting the building designated as a historic site was not successful, as the Historic Landmarks Preservation Commission of New York found the claim to be meritless. New York's urban planning system, by maintaining neutrality and by separating political from planning considerations, was not swayed by racial or political arguments. In September 2011, the Islamic centre quietly opened its doors.[32] Incidentally, this is an example of how more people's participation may not necessarily lead to the advancement of minorities' interests.

A similar opposition emerged against a proposed mosque on a four-acre lot in Temecula City in Riverside County, across the border from Los Angeles city. The opponents alleged that the mosque could be a "strategic foothold for extremists, and would undercut American values and laws." After many community meetings and on the recommendation of planning officials, the city council approved the project, holding that they "could only consider land use issues and not base their decision on religious, political and social issues."[33]

In chapter 4, I described the development of mosques, mandirs, and gurudawaras in Toronto, New York, and Los Angeles. The numbers in each city run into the hundreds and more are being developed every year. All in all, there are about 2000 mosques in the United States and more than 200 in Canada.[34] Most of these mosques, and other places of worship, develop as a matter of right within the existing zoning and planning policies. They stir up little or no controversy and are accepted like other community buildings. Usually, the opposition to non-Christian places of worship crystallizes around a newly built development in, or near, a long-established neighbourhood.

Of course there are cases where proposals for mosques, even churches, were fiercely resisted by local communities, resulting in the withdrawal or rejection of plans.[35] Usually, it is the political leadership of cities that is swayed by a strong opposing faction. Yet often, political leadership mediates solutions to help minorities meet their needs.[36] Planning organizations by and large keep the public discussion focused on matters of use compatibility, environmental and traffic impacts, and adequacy of services.

The history of the development of non-Christian places of worship unfolds as a process of mutual learning in which an emerging religious community learns local traditions and architectural idioms, while the local civic and political communities come to understand the rights and needs of minorities to establish their institutions. A study of mosque development in the Toronto area highlights this process of mutual learning. The proposals had to be modified in some cases and occasionally appealed before the province's Ontario Municipal Board against cities' decisions, but in all cases they were eventually success-fully completed.[37] By 2011, there were almost one hundred mosques in the Toronto area and none seriously proposed has remained undevel-oped, though some may have gone through long-drawn-out and costly review and appeal processes.

The evolution of mosque architecture in the United States and Can-ada recapitulates the process of social learning on the part of Muslim communities. From the Moorish design of the Islamic Cultural Center of Washington (1957), the mosque's functions and architecture evolved into the North American idiom, absorbing modernist/international forms and fusing them with historical Islamic styles and decorations. The Islamic Cultural Center of Manhattan (1991), designed by Skid-more, Owings & Merrill, Gulzar Haider's designs for the Islamic Center in Plainfield, Indiana (1981), and the Bait-ul-Islam mosque in Maple, Ontario (1992), are examples of the evolving North American Islamic style. In 2014, the Aga Khan contributed an international museum of Islamic art and an Ismaili Muslim centre to Toronto's expanding multi-cultural landscape.

The politically charged environment post-9/11 makes the develop-ment of mosques a test case of responsiveness to ethnic minorities. But the development of mandirs, Buddhist temples, and even Korean or Chinese churches has also been a process of social learning, both func-tionally and architecturally. Stacy Harwood cites the case of the Lien Hoa Buddhist temple in the city of Garden Grove, Southern California, which had almost a decade of dealing, negotiations, and a court case with the municipal planning department about alleged code violations by the temple. Yet the city council eventually negotiated a settlement, binding the head monk to some conditions for the building's religious use and permitted the building of a 578-square-foot assembly hall to facilitate its functioning.[38]

Many places of worship are established in existing stores, churches, or houses in commercial and industrial areas as a matter of permit-ted use under the prevailing zoning. It has happened in all cities. They are developed without much planning oversight as a zoned right. For example, along Coney Island Avenue in Brooklyn, New York, there are six storefront mosques over a stretch of about two miles. In the Toronto area, a number of mosques, mandirs, and churches have been established in malls, industrial parks, and storefronts. These developments do not suggest that there have been no public controversies, but that urban planning has responded to such developments as it would to any other permitted use.

To conclude the discussion of places of worship, a final point can be made. Community institutions are evolving and changing. Planning policies have to respond to new issues that arise with changes in the configurations of places of worship. Places of worship are growing big in size and are beginning to spawn faith-based schools, youth and senior centres, and halls for community gatherings. They draw around them residential developments, playgrounds, and health/recreational facilities for communities of belief and identity. Sandeep Agrawal describes four case studies, of a Hindu temple, Muslim mosque, Catholic church, and Sikh gurudawara in the Toronto area, each of which forms the focal point of a faith community and serves as a multifunctional place.[39] Some places of worship are proposed on large sites, of 10–20 acres, opening new lands for suburban development, contrary to planning goals of curbing sprawl and smart growth. Some are locating in industrial zones to avoid protests from local neighbourhoods about their parking and services impacts. These trends are raising broader policy issues of balancing cities' goals to preserve land and discourage sprawl and communities' desire to have multipurpose places of worship. Planning policies have to balance the area-wide public goals against ethnic communities' claims for equity and their apprehensions about discrimination. Not an easy balancing act.

The Politics of Land Use

Urban development is a politically charged process in which the real estate industry, local residents, businesses, community organizations, as well as in-migrating groups have a stake. Urban planners and political leadership are the targets of lobbying, arguments, protests, and threats of voter revenge. The politics of development can have an undercurrent of discrimination and racism if there are marked differences of class, race, and ethnicity between the old and new residents of a developing area. This is played out in battles for favourable land-use and zoning policies.

Mike Davis has traced the history of the politics of the Los Angeles region's development. The suburban communities' resistance to the movement of minorities and the working class into their neighbourhoods, or Nimbyism (not-in-my-backyard) was propelled by attempts to keep out Blacks, Latinos, and Chinese.[40] In the 1980s, for example, the city of Monterey Park had a nativist resistance against Chinese investors and in-coming households. It rallied around the flag of slow growth and controlled development. Yet as Chinese voters grew in number in the 1990s, the composition of the city council changed and the resistance collapsed.[41] By the 2000s, Monterey Park was a major Chinese city and Latinos were the ones being resisted. With cities becoming majority-minority, discrimination and racism are changing face. Interethnic conflicts are going to be the new expressions of racism.

The politics of development impinges on urban planning. Professional planners' codes of ethics bind them to the principle of social justice, serving the needs of the disadvantaged, promoting racial and economic integration, and respecting diversity.[42] These values are counterweights to the politics of discrimination. Yet planning decision making involves elected mayors and councils, community organizations, and property owners with constitutional rights. The pulls and tugs of their interests play out in the politics of city planning and development.

Racial and class inequalities in housing, employment, and the quality of neighbourhoods persist, largely because of the national and now global political economy. Witness the abandonment of large parts of Detroit, Toledo, Cleveland, and other industrial cities with the flight of manufacturing from these cities. Similarly, the mortgage-lending-induced financial crisis beginning in 2008 trapped large numbers of households in foreclosures, leaving large swathes of abandoned subdivisions in US cities. Stricter banking regulations saved Canadian cities from this fate. Obviously, these are causes of inequalities that bypass urban policies. Urban planning is also buffeted by the strong winds of national economic, technological, and ideological cycles.[43]

Cultural Divergences and Constructing an Intercultural Common Ground

A city has a thick weave of inter-linked land uses and interlocked activities, which mutually affect each other, both positively and negatively. The introduction of cultural differences into this mix injects divergences of social values and beliefs, precipitating potential land-use conflicts among neighbours. This is the arena of what Ash Amin calls "the daily negotiations of ethnic differences" and the "politics of micro-publics of social contact and encounters."[44] Frequently, these intercultural differences in the use of specific sites and neighbourly disagreements about particular activities end up at the door of urban planning. It addresses such competing interests by injecting public-interest criteria into negotiations. This is illustrated by the following examples.

The Chinese consider the presence of dead bodies near their homes and businesses to be a source of bad spirits. They do not want to live and work near funeral homes. Yet funeral homes historically are a low-impact (for parking, access, etc.) land use and are conventionally allowed in commercial zones. A number of proposals to build funeral homes in the suburbs of Toronto have run into strong opposition from the nearby Chinese residents. Urban planners tried to mediate the disputes, but found that existing zoning regulations allowed funeral homes as a right. Eventually, the zoning disputes ended up before the provincial appeal board, which invariably decided to uphold the zoning permission, explicitly maintaining that land use decisions cannot be based on "subjective" negative effects, and the compatibility of land uses must be decided on their physical and infrastructural impacts.[45] The planning system framed the decision within the parameters of its domain and authority.

Another common case of cultural clash is when mega-homes are built in established neighbourhoods of vintage houses and leafy lots. And if these mega-homes bring new ethnic groups, the design dispute can turns into a racial and ethnic confrontation. Variously called monster or mega-homes in Canada and McMansions in the United States, they often are permissible within the existing zoning envelope, except that the existing homes were not built to the maximum of the permissible volume. Hong Kong Chinese moving into the Kerrisdale and South Shaughnessy neighbourhoods of Vancouver ignited the resident English Canadians' protests. The City of Vancouver's planning department and the city council intervened and mediated between the two communities, coming up with new design guidelines (1993) requiring that new homes have regard for the streetscape and architectural compatibility.[46] Los Angeles's McMansions, often associated with rich Iranians, were regulated through special city ordinances in 2008 and 2010 requiring that they be consistent with the lot size, character of the neighbourhood, and steepness of slope.[47]

The neighbouring towns of Markham and Aurora in the Toronto area were petitioned by numerous Chinese homeowners for a change of their house numbers that had the digit four (4). Even some Anglo-Canadians wanted a change of house number, as they could not sell their homes in the neighbourhood evolving into a Chinese enclave. The digit 4 has the ring of the Chinese word for death and hence is considered unlucky. House numbers have implications for the sequencing of addresses for property identification, ease of deliveries, as well as identification by emergency services. Markham firmly refused to change house numbers, but Aurora passed a policy to allow a change if it does not disrupt sequencing, namely, if a house lot is on the corner and not in the middle of a block.[48]

Many such examples can be cited, but the point being argued is that highly localized and small-scale issues of land use and services are invested with cultural symbolism and values. Urban planning is called to negotiate this micro-politics of intercultural neighbourliness.

Lest this narrative give the impression that there is a veritable cultural war in multicultural cities, it should be said that it is tilted towards highlighting cultural differences as they play out in the competition for neighbourhood change and local development. A vast majority of people live harmoniously among others of different race, ethnicity, and beliefs, accommodating each other's needs and facilities.

Ethnic Enclaves, Commercial Malls, and Accommodations of Differences

Ethnic enclaves or neighbourhoods are the elephant in the room for multicultural planning. They are the most striking emblems of multiculturalism in the urban landscape. Yet urban planning has neither the mandate nor the tools to regulate the geographic distribution of people. It can affect enclaves' quality of life by providing facilities and services as well as through sign regulations, recognizing their special character, land use, housing, and commercial policies. However, planning policies cannot prescribe who may or may not live in an area. They cannot and should not zone areas by ethnicity or race quotas.

Ethnic enclaves are spatial expressions of people's choices for locations, homes, and neighbourhoods in the market. Of course, there are structural factors that define fields of residential choice for people; but exclusionary practices are not now usually the primary cause of enclaves. The strategic goal of urban planning, regarding enclaves, is to enhance the enclaves' economic and social vitality, reduce their segregation, and tie them closely to other parts of cities. The goal is to advance their social sustainability by increasing equitable access to the collective goods of a city and by integrating them into the web of city life. This is an ongoing exercise in neighbourhood planning, place making, and local economic and cultural development. But urban planning has developed neither comprehensive goals nor overall strategies for carrying out these various tasks. It has largely dealt with the challenges of ethnic enclaves in neutral but incremental ways.

The history of Chinese malls in Toronto is an example of accommodation and mutual learning. The first mall (1983), Dragon Centre in Scarborough, did not conform to the city's planning standards for commercial malls. Instead of a supermarket, it had a banquet hall as the anchor. Its stores were small and numerous for the floor area and its parking requirements exceeded. After public hearings and negotiations, the city suitably modified its requirements but the project's parking facilities were increased.[49] This established a precedent for the development of Chinese plazas and malls, including the Pacific Mall in Markham, one of the biggest Asian malls in North America. It is now turning into a node of Chinese commercial-professional activities with the expansion, redevelopment, and interlinking of the three existing Chinese mega-malls at the intersection of Kennedy and Steele Avenues in Markham and Toronto.

By 2013, there were sixty-six Chinese shopping centres in the Toronto CMA.[50] In Los Angeles, Koreatown was the path-breaking development that spawned ethnic-themed malls all across the United States. In Los Angeles city and the neighbouring municipalities, what James Rojas calls "Latino Urbanism" is transforming many neighbourhoods.[51] On the initiatives of Latino residents, rules are being loosened to allow the painting of murals on walls, street vending is gaining acceptance, and street festivals as well as pedestrian plazas are being incorporated into urban designs. Latino institutions and aesthetics in neighbourhoods are turning them into Mexican and Central American enclaves.

One mechanism used in urban planning to accommodate the cultural uniqueness of ethnic enclaves is to designate them as special districts, promoting local initiatives for appropriate zoning and design controls and community programs. Business Improvement Districts (BIDs) in the United States and Business Improvement Areas (BIAs) in Canada are planning programs that help set up local business associations in ethnic enclaves with the authority to levy charges on members and carry out local improvements and promotion.

There are a few planning strategies for the mixing of ethnicities. The most common is to increase the mix of types of housing to attract a variety of households in an enclave and to forge stronger links with other parts of a city. Another strategy is to increase social encounters by drawing others into enclaves through programs of art, culture, sports, and entertainment.[52] The point is that building intercultural bridges, geographic and social, is the strategy of integration.

The development of heritage art galleries and the opening of the Metro Gold Line station in Los Angeles's Chinatown is an example of the attempt to bring others into an ethnic community, as is the program to turn New York's Chinatown into an ethnic-theme commercial strip and build a Chinatown history museum as a point of tourist interest.[53] Such are the instruments of urban planning for integrating ethnic enclaves. Public places that draw people of diverse backgrounds are another instrument of promoting integration.

Ethnic enclaves present another challenge. They are seldom static, often changing. New households move in and the established ones shift to new places as they go through family cycles. Second generations of immigrants have different demands. They hang out at Starbucks and go to McDonald's as much as they visit ethnic restaurants. These evolutionary changes result in changing the structure of ethnic enclaves. For example, Chinese ethnic malls are beginning to include mainstream stores and fast-food chains.[54] They are turning into multi-ethnic shopping areas. Relatively older Chinese malls in Scarborough (Toronto) are drawing Korean, Japanese, and Filipino stores, as is the case with the Forest Hill Mall in New York. Obviously, urban planning policies have to be flexible to respond to evolving situations. Overall, urban planning has recognized the distinction of ethnic enclaves and attempts to accommodate as well as integrate them into the urban fabric, not in proactive but in reactive ways.

Social Deprivation and Ethnic Neighbourhoods

Although ethnic enclaves and neighbourhoods are generally viable communities of middle-class flavour, they can have clusters of poor households struggling to stay afloat. Such clusters, if large enough, turn into socially deprived areas where unemployment, crime, poor schools, and split families abound. In the United States, there are historical ghettos and minority neighbourhoods sustained by discrimination and public neglect. In Canadian cities, poor neighbourhoods are increasingly inhabited by immigrants and tend to consist of clusters of rental housing in inner suburbs.[55] Cultural differences are not the primary source of their deprivations, though race has some bearing on their long neglect. For urban planning, they present the age-old challenges of socio-spatial inequality compounded by racial and cultural differences.

Urban planning has emphasized improvements in physical conditions, infrastructure, and services to improve the living conditions in deprived neighbourhoods and historic ghettos. It has tried large-scale clearance and redevelopment of blighted neighbourhoods, initiated housing and infrastructure projects, and promoted economic development and community-services programs. It has experimented with various organizational models for the planning and implementation of slum improvement programs. They have ranged from public-sponsored mega-projects, to public-private consortiums and community-managed development and delivery of services.

These tools and lessons continue to be a part of urban planning's repertoire for slum improvement and overcoming social deprivation. One lesson that came out of the urban renewal programs of the 1960s and 1970s is that physical rebuilding alone is not enough for overcoming slum conditions. There have to be simultaneous efforts for social reconstruction and empowering people through the provision of social services and community organization. The physical and social-development programs have to be interlinked to unfold a process of comprehensive community renewal. Another lesson from the earlier programs is that the large-scale clearance of slums tears apart viable communities and displaces long-settled residents, and so has to be avoided. Physical blight is not to be treated by extensive surgery.

Yet neighbourhood improvements and gentrification in centrally located areas raise rents and lead to the gradual displacement of the poor. This is the policy dilemma of revitalizing deprived neighbourhoods: such programs improve living conditions but drive out poor residents. In market developments, the gentrification of neighbourhoods also leads to similar results.

From the perspective of responding to ethno-racial diversity, revitalizing programs are guided by the needs of the local population mix. If there are sizeable ethnic groups, provision is made to accommodate their cultural/religious needs, for instance by providing spaces for places of worship and offering culturally appropriate programs in community centres and facilities.

Another new strategy is to promote the mixing of low- and middle-income households and de-concentration of disadvantaged families living in poor neighbourhoods. Rent subsidies, housing vouchers, or homeownership tax credits are the tools used to enable poor families to access housing in middle-income areas, as is promoting infill developments of market housing and businesses to broaden the economic base of deprived neighbourhoods. The idea is to reduce the mutually reinforcing effects of concentrated poverty. This strategy evolved in the 1980s and 1990s, and continues to be applied in the 2000s. Some current examples of deprived-neighbourhood revitalization programs testify to their popularity as tools.

Regent Park, Toronto, is an example of the current revitalization strategies for deprived neighbourhoods largely inhabited by minorities and immigrants. It combines the efforts of public, private, and neighbourhood organizations. On a 70-acre site near the city's downtown that had 2083 public housing units exclusively, an imaginative strategy has been devised to rebuild in phases the entire community at higher density with mixed housing of both rent-geared-to-income and market-value types, complemented by commercial development and social reorganization.[56] By increasing density, an additional 3000 housing units will be built at market prices. Their revenue will fund the replacement of public housing for the resettling of displaced tenants. There will be new businesses and an aquatic centre as well as community-managed facilities for recreation and education. The first phase of the project is near completion. This example illustrates the strategy of mixing various social classes and diluting the concentration of socially disadvantaged families, all done with the participation of private developers and the local community, without net cost to the city. This approach of revitalizing ghettos and slums is spreading. Yet Regent Park's strategy cannot work everywhere. It is a publically owned site which is close to the downtown and in high demand. It is premised on generating funds from development that will help reinstall displaced residents in subsidized houses.

In the United States, private and community actors often are the primary driving forces. Cities and states help unlock their energies and resources. This is what has happened in Harlem, the mecca of Black community life in New York. It began to be gentrified in the 1990s, bringing in middle-class households, artists, and investors. That process itself was triggered by city and state investments in the redevelopment of 125th Street, the business and cultural hub of Harlem. The city has rezoned the area, increasing density and promoting the development of cultural and entertainment facilities. Major chain stores have opened and commercial activity is flourishing in Harlem. Former president Clinton maintains an office in the state tower on the street. Columbia University has assembled land for its expansion in West Harlem. Undoubtedly, these developments are not without controversy. There are local groups who see in these developments a threat to Blacks' control of their neighbourhood.[57] In response to the community opposition, the plan includes a $20-million affordable housing fund, generous relocation assistance for residents, two parks, and promises of hiring locals.[58] From the planning perspective, the point is that the development of mixed housing and mainstream businesses are the tools currently used to diversify the social and economic base of deprived ethno-racial communities.

New York and Toronto have used density bonuses to developers in return for their setting aside a quota of affordable units in their rental buildings. This policy helps mix social classes and ethno-racial groups.

Yet one need not be a Pollyanna about the promise of such strategies. The displacement and squeezing of minorities, particularly Blacks and Latinos in US cities and poor immigrants in Canada, continues to be the fallout from the public and private redevelopment of deprived neighbourhoods. Redevelopment occurs in spoonfuls rather than in the massive doses of the urban renewal programs.

Community-based social development in poor neighbourhoods is another strategy commonly used in deprived neighbourhoods and areas of ethno-racial concentrations. Toronto has a network of multi-service neighbourhood centres in Malvern, Thorncliffe, Fairlawn, Davenport, and other poor neighbourhoods of immigrants. These centres offer youth counselling, job search assistance, English language classes, seniors' clubs, settlement assistance, after-school programs, and so on. Leonie Sandercock refers to such centres as places for building intercultural communities, "where strangers become neighbours," and cites numerous examples from Europe, Australia, and Canada, including the well-publicized example of Vancouver's Collingwood Neighbourhood Centre.[59] The United States has a long history of community organization, including the settlement houses of the 1920s and 1930s, helping immigrants' resettlement. Altogether, the strategies for revitalization of deprived neighbourhoods are evolving with the experimentation of strategies and in response to changing political ideologies and economic conditions.

The limitations of public funds and private resources, exacerbated by years of budgetary cuts, also limit the effectiveness of such strategies.

The Strategy of Reasonable Accommodation

The foregoing narrative has illustrated how cultural differences call for adjustments and modifications of planning policies and the balancing of competing objectives. Most of these modifications occur reactively in response to a specific situation. Yet they fall into a pattern and that is the strategy of reasonable accommodation.

The strategy of reasonable accommodation is not frequently articulated in the urban planning literature, though it is recognized, even legislated in some sectors, as a principle of public policy. It is incorporated into labour law, disabilities legislation, and human rights codes in the United States and Canada.[60] To recall the discussion in chapter 2, reasonable accommodation is a two-pronged approach. It allows modifications in an institution's rules, requirements, and policies to accommodate the divergent needs and demands of clients and staff, but sets limits of reasonableness on such accommodations so that they should not compromise the institution's objectives, cause undue hardship, impose excessive costs, and have adverse effects on others.[61] Despite some reservations about the concept from both the right and the left,[62] reasonable accommodation is practised widely because it responds to diversity while sustaining institutional integrity.

The practice of urban planning has a long history of allowing minor variances from zoning and design regulations on the basis of differences in site configurations and functional needs. This practice is nothing but reasonable accommodation. Similar accommodations for particular conditions abound in housing policies, community services, environmental regulations, and physical planning. A policy of reasonable accommodation on the basis of cultural and racial diversity is not a big innovation. Its two-pronged approach is particularly relevant for urban planning, which aims to balance the cultural needs of ethno-racial groups with the common interests of civic society.

Policy Index of Socio-cultural Accommodations

Within the institutional parameters of urban planning, a set of policies can be conceived to respond to cultural diversity. Together, they offer a checklist of measures that recognize ethno-racial differences and accommodate them in planning policies. One such attempt is reflected in the "policy index of multicultural policies,"[63] which includes guidelines and procedures for accommodating cultural differences in both the process and substantive aspects of urban planning. It is a list of eighteen planning policies and activities that range from modes of involving diverse communities in planning decision making to establishing culturally responsive policies of land use, housing, community services, and business development. It must be noted that these policies are meant to modify and tweak the criteria and programs of institutionalized urban planning. They represent an approach of infusing pluralism in urban planning institutions. They do not suggest a parallel system of planning for ethno-cultural communities, but aim at incorporating their interests into existing practices. The policy index is displayed in the box on the next page.

In practice, accommodations are requested and granted on a case-by-case basis. Yet there is an element of arbitrariness in leaving those to be worked out individually. They have to be turned into rules and criteria to make outcomes in similar cases consistent. This is the process that Gerard Bouchard and Charles Taylor call the harmonization of concerted accommodations.[64] Harmonization practices are a step towards reconstruction of the common ground, which lays the basis for accommodations of cultural differences. In physical planning it means, for example, establishing criteria for recognizing and expressing cultural differences through land-use policies and zoning by-laws, while devising performance standards that allow for flexibility of design. Other examples of harmonization are citywide policies for linguistic interpretation, and for the sustainability and integration of ethnic enclaves, programs to promote and regulate ethnic economies, public health regulations for diverse lifestyles, the conservation of minorities' heritage, art and culture facilities for ethno-racial groups, and rules for non-English signs. These measures define the scope of responsiveness to cultural differences. They lay the bases for multicultural planning. The policy index is a checklist of accommodations that can be embedded in planning policies.

The policy index has been used to assess the multicultural planning practices in selected cities of Canada and the United States. A mailed questionnaire based on the policy index was sent to the planning departments in the two countries in 2008–9.[65] Forty-two cities responded, including municipalities in the metropolitan areas of Toronto, Los Angeles, Chicago, San Francisco, Houston, Montreal, and

Vancouver. Among the responding cities were twenty-three American and nineteen Canadian municipalities of these urban areas, including many suburban jurisdictions. This survey gives an overview of the state of multicultural planning practice in North America.

Policy Index of Multicultural Planning

1. Providing minority language facilities, translations, and interpretation in public consultations.
2. Including minority representatives in planning committees and task forces as well as diversifying staff.
3. Including ethnic/minority community organizations in the planning decision-making processes.
4. Routinely analysing ethnic and racial variables in planning analysis.
5. Studies of ethnic enclaves and neighbourhoods in transition.
6. Recognition of ethnic diversity as a planning goal in official/comprehensive plans.
7. Citywide policies for culture-specific institutions in plans, e.g., places of worship, ethnic seniors' homes, cultural institutions, funeral homes, and fairs.
8. Policies/design guidelines for sustaining ethnic neighbourhoods.
9. Policies/strategies for ethnic commercial areas, malls, and business improvement areas.
10. Incorporating culture/religion as an acceptable reason for site-specific accommodations/minor variances. Accommodation of ethnic signage, street names, and symbols.
11. Policies for ethnic-specific service needs.
12. Policies for immigrants' special service needs.
13. Policies/projects for ethnic heritage preservation.
14. Guidelines for housing to suit diverse groups.
15. Development strategies taking account of inter-cultural needs.
16. Promoting and systemizing ethnic entrepreneurship for economic development.
17. Policies/strategies for ethnic art and cultural services.
18. Accommodating ethnic sports (e.g., cricket, bocce) in playfield design and programing.

(Source: Mohammad Qadeer, "What Is This Thing Called Multicultural Planning?" *Plan Canada*, 2009, 13)

The State of Culturally Sensitive Planning

The urban planning process has the difficult task of balancing divergent social and cultural needs, on the one hand, and reconciling them with public goals that affect everybody, on the other. Within these parameters, urban planning practice has been reasonably responsive to cultural differences. It is borne out by the survey of planning departments referred to above. The survey shows that most of the responding cities have policies that lay the bases for accommodating cultural differences.

The patterns of adopted policies suggested by the survey (table 10.1) include:[66] (1) Canadian large and medium-size cities have more culturally responsive policies than do the US cities. (2) Out of 19 possible policies, the large US cities have 12.6 policies – about two-third of the total – but Canadian multiculturalism pushes this percentage even higher, to more than four-fifths. (3) Canadian-American differences are statistically significant. (4) The number of adopted policies generally fall with the decreasing size of cities, the major dividing line being between the large cities and the rest. Obviously, large cities have sizeable ethno-racial groups and they demand accommodations as fair treatment. (5) The small cities in this study were all suburban municipalities within metropolitan areas. Their low number of adopted policies, especially the US cities' numbers being a bit higher than the Canadian cities', are indicative perhaps of their limited planning authority and dependence on regional policies.

A look at the policy index, on which the survey was based, shows that policies fall into three groups, relating to (1) in-puts in planning decision processes, (2) land use and development matters, and (3) the provision of community services. The survey shows that cities of all sizes were following all the listed culturally sensitive policies about planning process, such as facilitating minorities' involvement in planning, followed by policies bearing on the provision of culturally relevant community services, with the policies for accommodations in land use and development being the lowest in incidence.[67] These differences in the incidence of policies in the three aspects of urban planning can be explained by the fact that the planning process and community services fall largely in the public sphere, whereas the initiative for physical development begins with the private sector.

The survey also revealed shortfalls in the harmonization of culturally responsive policies. None of the forty-two cities had established comprehensive policies for the management of ethnic enclaves, businesses,

Table 10.1 Mean number of adopted policies, American and Canadian cities

	Large cities, 500,000 and more	Medium-size cities, 100,000–500,000	Small cities, fewer than 100,000
Canadian cities	15.4	11.6	5.0
US cities	12.6	6.5	7.0

Source: Mohammad Qadeer and Sandeep Agrawal, "The Practice of Multicultural Planning in American and Canadian Cities," *Canadian Journal of Urban Research* 20, no. 1 (2011), 144, figure-2.

and cultural/religious institutions.[68] They have not developed overall criteria to guide accommodations of cultural differences. Toronto, New York, and Los Angeles have motherhood statements about promoting and valuing diversity in their comprehensive/official plans for future development. Yet they have no specific policy or measure in the plans that link actions to this goal. Accommodations are made on a case-by-case basis.

Despite the lack of comprehensive approach, the practice of multicultural planning through accommodations appears to be widespread. It is ahead of planning theory. It has grafted multicultural forms and functions on the existing institutions of cities. Yet the proactive approach to multicultural planning, in the form of clearly defined objectives to guide the development of the culturally diverse spatial organization and establishing criteria for reasonable accommodations, is lagging. There is no articulated vision and relatively few area-wide policies of multicultural planning.

Multicultural Planning in Civil Society

I have discussed the infusion of cultural responsiveness into the theory and practice of institutionalized urban planning, which is mostly carried out in the planning departments of cities and regions. This process also occurs in non-governmental agencies. Community organizations, interest groups, and neighbourhood associations, for example, develop programs to serve diverse communities in culturally sensitive ways. This is another form of multicultural planning based on the initiatives of a civil society. It is the phenomenon of community-based planning.

Multicultural planning in the non-governmental sector occurs in four types of settings.

1. Ethno-racial advocacy groups that lobby for equality and non-discrimination for the fulfilment of social and cultural needs of their respective communities, for example, the NAACP and B'nai B'rith.
2. Community organizations that plan and provide a service for a particular group, such as a Jewish seniors' housing agency or South Asian women's centre. Social agencies promoting and providing multiple services for an area include the United Way Toronto and West Harlem Community Organization.
3. Neighbourhood associations or centres for the multifunctional development of an ethno-racial or immigrant group, for example, the Greater Chinatown Community Association in New York and Thorncliffe Neighbourhood Centre in Toronto.
4. City- or region-wide agency for social planning and development, for example, the Community Social Planning Council of Toronto and Black United Fund of New York.

Non-governmental multicultural planning is particularly relevant in the advocacy of equal treatment for minorities and in highlighting hidden ethno-racial biases in public policies. It is also an effective mode of organizing social services for targeted racial, cultural, or geographic communities. On the process side of urban planning, it brings multicultural perspectives into the decision making. Yet the non-governmental mode of urban planning has limitations.

It is partial in coverage and limited in function. Most of the non-governmental initiatives in urban planning are focused on the neighbourhood level and not on a city as a whole. Many organizations concentrate on a particular community and either advocate for their interests or cater to their needs (for example, the Chicano Resource Center, Los Angeles). The point is that the institutionalization of pluralism and laying the ground for the cultural harmonization of policies can take place primarily in the public realm. Every multicultural city has hundreds, if not thousands, of community organizations and associations, but their efforts do not add up to a strategy of multicultural planning at the city level.[69] That is only possible in institutionalized urban planning.

Multicultural Planning and the Public Interest

Toronto's official plan (2002/10) has as its first guiding principle "diversity and opportunity," followed by beauty, connectivity, leadership, and stewardship. Yet after this motherhood statement, it goes on to spell out citywide policies about increasing density, compact development, mixed neighbourhoods, the preservation and enhancing of river valleys, increasing public transit, and building a sustainable and vibrant city.[70] How diversity will be promoted is not linked to any policy or strategy statement in the rest of the document. This is not an unusual situation. New York's general plan outline, drafted as part of its effort to update zoning by-laws in the mid-2000s, has a similar statement about racial and ethnic diversity and equity. Diversity is recognized as a value to be promoted, but urban planning largely proceeds on the basis of the common good, concentrating on policies that focus on functional elements of a city, such as policies to regulate density and land use by zoning, promote environmental and infrastructural sustainability, and develop citywide transportation and housing strategies. Urban planning is driven by civic goals and public interest by manipulating and guiding the city's structural and functional elements.

Ethno-racial differences of culture and needs are dealt with as accommodations in the functional-structural policies. They may lead to some revisions and harmonization of functional policies (such as revising church standards into generic places-of-worship policies), but they do not override the functional focus and public-interest criteria. Undoubtedly, functional policies are coded in the mainstream culture, but ethno-racial groups share and own many of those historical/universal values and norms. For example, the promotion of public transit and reduction of automobile gridlock are policies that almost all ethno-racial groups will subscribe to. For them, equity means that they are equal beneficiaries of such policies. Their cultural differences are accommodated and harmonized within the scope of shared objectives and the common good. Shared citizenship and rights to the city flowing therefrom circumscribe the domain of multicultural planning. It consists of embedding ethno-racial interests in the long-established framework of urban planning's policies and standards, giving rise to some form of pluralism in some policies. It is not a matter of developing different plans and policies for different groups.

Conclusion

Multicultural cities. Toronto, New York, and Los Angeles, and most others, have made accommodations to cultural differences in their urban planning policies and programs. These accommodations have resulted in the incorporation of ethnic enclaves, places of worship, and commercial establishments, multilingual access to city services and signage, and museums and a variety of cultural facilities into urban structure and institutions. The patterns of accommodation and subsequent harmonization of polices are similar in the three cities, despite differences in the political ethos and constitutional laws and practices between the United States and Canada. With a greater emphasis on private rights, the US cities seem to need and offer slightly fewer accommodations than the Canadian cities. Yet the differences are made up for by the ethno-racial communities' greater leeway in fulfilling their needs. This chapter has offered many illustrative examples of the incorporation of differences in both countries.

The formation of pluralistic policies follows the same path as has been witnessed in other aspects of the formation of multicultural cities. There are long-established existing urban structures and institutions in a city, into which are infused culturally and racially differentiated norms and practices. A pluralistic ethos incorporated in urban institutions makes them multicultural, while retaining the common ground of shared space, infrastructure, laws, traditions, and civic culture.

The functional elements of mainstream institutions are the focus of planning in multicultural cities, for instance, land use, transportation, services, and utilities. Many of their objectives are part of the shared citizenship. Equity in the distribution of their benefits signals the success of multicultural planning. Mainstream institutions, appropriately modified by reasonable accommodations, are the instruments of public interest and common good that ensure a sustainable city and good quality of life for all, including ethno-racial groups. They are the common ground in which pluralistic policies take root. Multicultural policies in a city do not result in different policies for different groups, but promise responsiveness to cultural and social differences, while maintaining the structural-functional unity of the city. Multicultural responsiveness is one among the many objectives of urban planning.

Planning practice in multicultural cities has forged ahead of planning theories and discourse. Of course, many inequities in the distribution of housing and community services persist, particularly along racial lines. These inequities are reinforced by poverty and economic inequality, matters on which city policies have a very limited impact.

Canadian and American forms of multicultural planning differ in process. The former is centred on legal-bureaucratic institutions, whereas the latter plays out mostly in the political arena, with greater private freedom.

The shortfall of the current practice of multicultural planning is its lack of a vision of the multicultural city. None of the surveyed cities have harmonized policies for ethnic enclaves or the role of ethnic businesses in the local economy, guidelines for programs of culturally differenti- ated facilities and services, or design standards for ethnic architecture. These are examples of issues that remain to be addressed in policies for multicultural planning. The policy index cited in this chapter is an example of the protocol that multicultural planning has to develop. On top of all these considerations, cultures change over time and so will the policies for multicultural planning.

Chapter Eleven

Imagining Multicultural Cities

Making Cities Multicultural

The ethos of diversity infused into a city's structure and institutions makes it a multicultural city. What does the ethos of diversity mean in concrete terms in a city's life? It means that people of different races, faiths, and cultures live as equal citizens and that their differences are reflected on streets, in public places, in neighbourhoods, and in everyday life. This book has documented how the recognition and incorporation of diversity embeds pluralism in the spatial organization and the economic, social, political, and policymaking institutions of multicultural cities. The evidence for this transformation is drawn from analytical observations of the accommodations of cultural differences in the structures and institutions of the three acknowledged multicultural cities in Canada and the United States, namely, Toronto, New York, and Los Angeles. From the experiences of these three cities, in two countries of varying cultural ideology, I have attempted to identify the characteristics that have made them vibrant multicultural cities. This inquiry has sought to identify the defining elements and conditions of a multicultural city.

A city with a long history and established institutions exists before it becomes multicultural. This city serves as the base over which pluralism and diversity are imprinted. Its internal structure, landscape, environment, economic organization, facilities, and services as well as governance and civic institutions have been set in place for decades and centuries. They are the medium for infusing differences and simultaneously bonding different groups into a community of citizens. They define both the scope and limits of incorporating diversity. The book has used this two-sided model of multiculturalism in analysing the incorporation of diversity into the structures and institutions of the three cities.

Often a multi-ethnic and multiracial population is taken to be the indicator of a city's multiculturalism. Yet diversity of population alone does not make a city multicultural. Ethno-racial diversity is a necessary but not sufficient condition for multiracial cities. By guaranteeing fundamental freedoms and respecting cultural and racial identities, the regime of civil and human rights transforms a multi-ethnic population into multicultural citizens. For example, Jeddah, Dubai, and Singapore are multi-ethnic and multiracial cities, where people have come from all parts of the world to work. Yet non-native residents have limited social rights, and even their residential visas to stay have to be periodically renewed. These are not multicultural cities, despite their cosmopolitan population. Toronto, New York, and Los Angeles are multicultural cities for recognizing and accommodating cultural differences as a matter of civil rights.

Human and civil rights entitle individuals and groups to the recognition of their identities and reasonable accommodation of the diverse ways in which their needs are met. They lay the foundations of multiculturalism by extending equality on cultural and racial bases. Equality rights build up the pressure to make a city's government, politics, markets, and civil society responsive to cultural diversity. They inject pluralism into urban institutions – pluralism that not only engages with diversity, but also seeks understanding across cultural lines and does not fall into a morass of relativism but is based on dialogue and encounters.[1] This is how a city becomes a multicultural city.

A multicultural city is not a separate species of city. It is an existing city with a new configuration of its institutions, ideology, and consciousness. It is a place where a diversity of physical forms, social practices, and cultural customs thrive as rights, where pluralism is grafted onto institutions. The incorporation of diversity results in the development of some common characteristics in multicultural cities.

The Common Elements of Multicultural Cities

Demographic Diversity

First, an obvious but defining quality of multicultural cities is the racial and ethnic diversity of their populations. Toronto, New York, and Los Angeles are internally so diverse that no one group numerically dominates. All three have either become majority-minority cities or are at the cusp of becoming that, so much so that their native-born Whites are now minorities, less than 50% of the population. In Canada and the United States, this is becoming characteristic of most metropolitan cities. Except for the linguistic convention, one cannot demographically call ethno-racial groups minorities. This demographic change is transformative.

As the ethnic and racial mix of the urban population changes, consumers' demands and preferences shift towards culturally preferred goods and services – for example, ethnic groceries, restaurants, apparel, music, art, furniture, and book stores of distinct cultural/religious markers are started – changing the commercial structure of a city. Though this is largely a market phenomenon, cities' receptivity to such developments has to be deliberately worked out through accommodations and the harmonization of economic policies, zoning by-laws, and signage regulations among many other public measures.

Realigning of Social Geography

Second, diversity realigns the social geography of cities. It transforms old neighbourhoods by bringing in new residents of a distinct ethnicity to replace the inhabitants who are vacating. Immigrants who are relatively younger than the aging native-born are the obvious replacement for the falling population in old neighbourhoods and are also the major buyers of new homes. The concentration of a particular ethnic group and the development of its commercial, religious, and community institutions turn an area into an enclave. The three cities are now dotted with ethnic enclaves.

The Toronto metropolitan area shows a broad sectoral alignment of ethnic enclaves, Chinese in the north-east, South Asians in the northwest, and Italian and Jewish neighbourhoods in the middle, all spreading towards the outer suburbs.

Los Angeles County and the city present a similar sectoral pattern, though it has historical roots in the spatial divisions of the city: namely, Blacks in South LA, Anglos in the west, and Latinos in the east. The new ethnic enclaves are following the same lines for advance outwards, except for the Asians, particularly Chinese, who have settled in a string of suburban municipalities of San Gabriel Valley.

New York's new ethnic enclaves have also appeared in outer boroughs, particularly in Queens and Brooklyn. New Chinatowns have emerged in Forest Hill and Sunset Park. Colombians, Mexicans, and South Asians form distinct clusters along Roosevelt Avenue in Queens. Joseph Berger lists sixteen ethnic neighbourhoods in New York, most in outer boroughs. Chapter 4 gives a detailed account of the enclaves in the three cities.

The multicultural developments in the three cities point out some common features of ethnic enclaves. (1) Ethnic enclaves of the post-1965 immigrants and their descendants are largely middle class and suburban, with homeowners for the most part and pockets of poor ethnics in rental apartments. (2) These are vibrant places combining ethnics' residential concentration with their commercial, religious, and service establishments. (3) Enclaves are neither the result of social exclusion nor do they overtly exclude others. (4) Usually a particular ethnic group is dominant, but small minorities of others are also present. (5) Enclaves are woven into the fabric of a city by threads of employment, leisure, commerce, infrastructure and civic services, education, and politics. (6) They are expressions of the ubiquitous urban process of clustering similar and complementary activities.

In the three cities, there are also vast areas of mixed neighbourhoods, sometimes called global neighbourhoods, where people of divergent backgrounds live side by side. A majority of the population, including ethnics, lives in such neighbourhoods. There is no part of these cities where the presence of "others" is not visible. Their presence is symbolized by residents, stores, restaurants, symbols and signage, workers and visitors, and languages heard on the street. For example, almost every part of multicultural cities has Chinese, Indian, or Mexican restaurants.

Cumulatively, these developments represent the infusion of cultural diversity into the urban spatial structure. The social geography of multicultural cities has an overlay of culturally differentiated residential and commercial clusters on top of its class and functional contours. Its ethnic landscape is built largely by private initiatives, but made possible with a public "buy-in" of the diversity through accommodations of differences and the recognition of cultural rights.

One last point about the spatial structure of multicultural cities is that its cultural contours are engraved over a long-established land and infrastructural system, which defines the scope and limits of diversity. It is the foundation of everybody's liveability.

The Ethnic Contours of Economies

Third, another manifestation of multiculturalism is the emergence of ethnic and racial sub-economies in the economic organization of cities. As chapter 5 documents, in the three cities, ethnics' and immigrants' entrepreneurship has resulted in the specialization of particular ethnic groups in specific economic activities. They start new lines of business and dominate in many existing activities, combining their human and financial resources with social and cultural capital, and forming ethnic economic niches and clusters.

Among the examples are Toronto's airport taxi-limousine service as a niche of South Asians, the Chinese economy of ethnic malls and domination of the computer hardware industry, and Italians' concentration in construction. New York has Chinese niches in garment manufacturing and the knock-offs trade as well as a Jewish concentration in medicine. Los Angeles's Chinese economy of computer and electronic businesses is a platform for the Pacific trade. Garden centres, nurseries, and construction are Latino-controlled economies. Ethnic economies increasingly are becoming a link in the chain of global trade. They are not just serving immigrants' commercial networks. Ethnic economies are backed by ethnic social networks, which span many countries.

There are remarkable similarities in the specialization of economic activities by ethnicity in the three cities. Yet not all ethnic groups are equally successful as entrepreneurs and in forging economic niches. The earnings of non-White and foreign-born labour are low. Ethnic economies provide jobs to new immigrants, but they can also turn into a trap for them. Ethnic entrepreneurship does not reduce inequalities within and among ethnic communities, but it creates opportunities for immigrants. All in all, ethnic economies and niches are subsystems of regional economies. They represent another form of the infusion of diversity into cities' institutions.

Community Structures and Social Organization

Fourth, ethnicity and race have played a large role in the social organization and community structures of cities. They are the basis of community networks and neighbourhood formation. Social relations in cities are organized around shared values, customs, language, religion, and interests. Ethnicity and race encompass many of these interests and are the fonts of identity. People may live in mixed neighbourhoods, but still have their relatives and friends spread all over a city forming communities without propinquity. The three cities show many parallels in the formation of network communities, though the literature's focus on spatial segregation tends to overemphasize the territorial bases of relations among races, ethnicities, and classes. But territorial proximity is only a small part of the social relations in cities.

Chapter 6 shows parallel institutions and processes of community formation in the three cities. It shows that race matters. Blacks and dark-skinned Latinos (Dominicans in New York) are the most segregated, particularly in the two American cities. Toronto does not have a historic native Black community of notable size, whereas New York and Los Angeles have large neighbourhoods of poor Blacks going back to the 1930s and 1940s. Spatially, the most segregated group in Toronto are Jews, though politically, economically, and socially they are well integrated. Their situation underlines the limitation of viewing spatial concentration as an indicator of social segregation.

Overall, Asians are more concentrated territorially in the three cities compared to the native Anglos. Yet there are variations by nationalities within these broad categories of ethnicity. The Japanese are fairly dispersed, whereas relatively prosperous Indians and Iranians are spatially more concentrated in Los Angeles. These examples illustrate the specificity of the social patterns of each group.

Urban social organization has overlapping networks. The civic culture and social interests cut across ethnic and religious lines. There are professional bodies, clubs and associations, art and culture groups, sports and festivals, reading circles, political parties, and other such cross-cutting social networks that overlay ethnic and racial bonds. The three cities so far show a fairly good level of social bridging through these cross-cutting networks.

Public Culture

Fifth, ethno-cultural diversity injects new activities, interests, and aesthetics into the public culture of cities. The music of distant lands reverberate on radio and television, in parks and streets. Festivals, fairs, parades, and demonstrations in support of, or opposition to, international causes become almost a daily occurrence somewhere in multicultural cities. New sports are introduced and parks and arenas are modified to accommodate them, for instance, cricket pitches and Bocce lawns in city parks. Television, radio, and cinemas offer the fare of many cultures and languages. Robust local ethnic media, both print and electronic, have emerged in the three cities, in addition to satellite TV and the Internet, which bring programs from all over the world into North American homes. Chapters 7 and 9 document the growth of such services in the three cities.

For example, Toronto, New York, and Los Angeles have TV channels offering 24-hour programs and news in Spanish, Chinese, Hindi, or Polish. Toronto has almost three hundred ethnic newspapers, sixteen in the Urdu language alone. New York has ninety Chinese, six Spanish, and three Polish newspapers and magazines, to cite a sample of the scope of the ethnic media. Bollywood and Hollywood are collaborating to produce a new genre of fusion movies in Los Angeles. The point is that everyday life and civic institutions have opened up to new forms and activities with the inclusion of ethnic communities.

The three cities have responded to cultural diversity by offering grants to community groups, supporting ethnic organizations, establishing offices for immigrant services, and legislating pluralistic policies. City libraries have been at the forefront, adding ethnic materials to their holdings and offering English-language/cultural classes to draw in new immigrants. The mainstream museums and concert halls have expanded their repertoires to offer the art and music of different countries. This infusion of diversity into cultural institutions has made multicultural cities places where variety rules and the fusion of art, music, and aesthetics spurs creativity.

A note of caution is in order here. I do not mean that the process of cultural infusion has been without resistance or setbacks. Almost every day, in some part of the three cities, there is some cultural/religious controversy. Charges of racism fly and disparities between the mainstream cultural institutions and fledgling ethnic art and culture are highlighted. Police almost everywhere are accused of targeting non-Whites. Yet mutual recognition prevails, and adapting to cultural diversity remains an ongoing project.

Pluralism and Public Policies

Sixth, the path for incorporating diversity in cities' structures and institutions is laid by public policies and political responsiveness to residents' rights. The institutional change towards multiculturalism has not come from a comprehensive strategy in the three cities. In their different ways, all have responded to cultural differences in incremental steps by enforcing citizens' rights. School boards have made their curricula inclusive and accommodated language differences by offering support for English-as-a-second-language students. Health and social services are adopting cultural competence as one of the criteria for dealing with diverse clients. All three cities have instituted translation and interpretation support services in making city programs accessible to non-English speakers. Their city planning departments have suitably made accommodations in zoning and other regulations for ethnic and religious activities. The New York police department has compiled a booklet of the cultural practices of different groups to sensitize its officers. Toronto's police are actively recruiting minorities. The point is that multicultural policies are adopted as problems develop and the necessity of recognizing cultural differences is realized to fulfil the professional objectives of city departments and organizations. Yes, gaps remain, but the process of institutional change has been laid out.

The voting power of immigrants, as they become citizens, has necessitated that political parties and elected leaders recognize their needs and court them with responsive policies. Ethnic community organizations have mobilized civil societies in the three cities to agitate for fair treatment and equality. In this regard, the American Blacks' civil rights struggle has opened the doors for immigrants and ethnics. Finally, the Canadian Charter of Rights and American Civil Rights Act have sustained the ethno-racial groups' drive for fairness and equality. All these strands of political, social, and legal processes have helped transform cities.

Although the three cities have liberal-democratic values and similar institutions of governance, they combine these elements in distinct configurations. The framing of multiculturalism in the local context has resulted in the emergence of similar patterns, but of varying forms, meanings, and spirit. The multicultural configuration of each city has a distinct flavour.

Multicultural Toronto

The Toronto Census Metropolitan Area of 5.82 million people includes sixteen municipalities and three second-tier suburban regional municipalities responsible for transportation, police, and social services as well as infrastructure on an area-wide basis. Toronto city, the heart of the metropolitan area, is the financial capital of Canada. It is a clean, orderly city of English heritage, where Jews, Italians, and other Southern European immigrants make for its mosaic. The multi-ethnicity of the area exploded after the removal of quotas for non-European immigrants in 1965.

Canada explicitly subscribes to the ideology of multiculturalism and has a legislated policy promoting it. Its Charter of Rights and Freedoms (1982) is the capstone of Canadian constitutional civil rights. The Charter is the primary basis of litigating for equality rights. Canada's multiculturalism policy gives ideological recognition to diversity and promotes ethnic cultures, but is seldom relied upon to adjudicate cultural rights.

In the previous section, Toronto's enclaves, ethnic malls, and social, art, and cultural pluralism were sketched out to illustrate how multiculturalism has been infused into almost all aspects of the city's life. Pico Iyer, called by the *Literary Quarterly* "the poet laureate of wanderlust," in discussing Toronto says it "has embarked upon multicultural experiment with itself as guinea pig."[2]

The distinguishing part of Toronto's openness to multiculturalism is that it is embedded in the Canadian tradition of orderly social change rooted in the constitutional values of peace, order, and good government. The process of institutional change is often initiated by ethnics' demands, but public responsiveness to accommodate their needs and harmonize policies has made the process orderly. Toronto's multiculturalism has the feel of a top-down public infusion of diversity.

The Toronto area takes pride in its multiculturalism. It has come to define itself by that. For example, the official motto for Toronto city is "Diversity Our Strength." This discourse defines public institutions' receptivity to pluralistic policies and ethno-racial groups' right to the city.

Other factors that have helped shape multicultural Toronto are (1) its polycentric urban form with a dominant and dynamic nucleus in the downtown of the central city; (2) non-party-based municipal politics and a political culture of responsiveness to civil society and business interests, with strong provincial oversight; (3) enforcement of the Canadian Charter of Rights and Freedoms through legislation, policies, and the courts, including the Ontario Human Rights Commission; (4) the absence of historic racial ghettos, and relatively good housing stock and viable neighbourhoods; and (5) a strong social welfare state, with universal health and educational services, income support, and immigrants settlement services – though in an era of governmental cutbacks many of these services are being reduced. These are the elements of the "local context" that invests Toronto's multiculturalism with order and coherence.

Myer Siemiatycki, Tim Rees, and Khan Rahi observe that "Toronto has proven adept at capitalizing on its diversity."[3] It has progressively harmonized its policies and practices to institutionalize cultural sensitivity. Urban institutions have been realigned but not displaced. The example of its social geography is a concrete illustration of this approach.

Yet the smoothness of Toronto's cultural transformation does not mean that ethno-racial disparities have been eliminated. The Toronto area is not without its challenges of integration. It has protests about racial profiling and public complaints about employment inequities. Despite these challenges, Toronto city is held to be a world-class example of multiculturalism.

Cosmopolitan New York

The United States and particularly its big cities are multicultural in the full sense of the term, but the "melting pot" national myth has influence over the popular narrative. Undoubtedly, the United States has a strong national identity and an assertive American creed, but its sub-cultural diversity and racial as well as ethnic differences are woven in its history and culture. Scholars have long recognized the limits of assimilation and the persistence of racial and cultural identities. In academic circles the multiculturalism of cities is widely acknowledged, so much so that Nathan Glazer, the doyen of ethnic studies, declares, "We are all multiculturalists now."[4] Yet in New York, and the United States in general, the public discourse is framed by race and ethnic relations and the term multiculturalism is heard only sparingly in popular narratives. Race in particular is at the forefront of the American discourse.

New York, a city of 8.4 million people, has always been the home of immigrants, whose successive waves landed on its shores to weave their identity into its landscape and social structure. Consciousness of ethnicity and race is woven into New York's civic image. In the pre-civil rights days, there were ethnic neighbourhoods and racial ghettos. WASP privilege was accepted. It was not a multicultural city, though it had a rich mosaic of cultures and races.

Multiculturalism has been brought about by the enactment of civil rights and by Blacks blazing the trail for the recognition of their identity. Poet and novelist Ishmel Reed says, "There would be no multiculturalism without Black arts."[5] Nancy Foner concludes that New York's history of immigration, its political structure, and its culture have "implications for the dynamics of ethno-racial identities and relations."[6]

The multiculturalism of New York bears many similarities to the institutional and structural expressions of diversity in the Toronto area. These common elements have been discussed above. The differences in its multiculturalism lie in its pragmatism and greater reliance on private initiatives. New York's multicultural forms are flexible, and have multiple origins. Diversity is enacted readily through private actions, which have much greater latitude. To illustrate this point, the example of unique private businesses in Queens is striking. They offer services to ethnics and immigrants "to clean their accent." In Toronto, such a need may be strongly felt, but the narrative of multiculturalism will shy away from acknowledging barriers of accents in communication.

New York has been a global city since well before the term was coined. It has had the headquarters of the United Nations since 1945. People from all parts of the world converge on New York for international negotiations, business, and fun. It is a world city that has thrived on accepting other cultures. It has an energetic, assertive, and liberal civic spirit that pervades ethnic cultures. The politics in New York is organized around party machines, lobby groups, political clubs, and, lately, community organizations. Coalition building and political bargaining are common modes of operation. This political culture enables ethno-racial groups to negotiate for their interests in political arenas. It is not a surprise that minority representation in the new city council is almost near 50%, the highest among the three cities. Philip Kasinitz, John Mollenkopf, and Mary Waters observe that New York has "a new kind of multiculturalism ... of hybrids and fluid exchanges across group boundaries."[7]

There are other specific conditions of New York's urban structure and political-social institutions that breed its open, negotiated, and fluid form of multiculturalism. John Mollenkopf, in theorizing about New York's differences from Los Angeles and Chicago as urban models, identifies the following attributes: (1) its density, strong city centre, and extensive public transportation system bring people of all backgrounds into close contact and sustain identities; (2) generous public services contribute to the social sustainability of immigrants; (3) as a global tourist destination it cultivates a cosmopolitan civic culture; (4) the city government and political system have a tradition of public responsiveness; and (5) the immensity of its economy and elaborate division of labour promote a variety of "small worlds" (economic niches).[8]

Post-modern Los Angeles

The city of angels, Los Angeles, is a spread-out city of 3.8 million people (2010), but its census metropolitan statistical area extends over Los Angeles and Orange counties. In this book, I have limited the analysis of metropolitan characteristics to Los Angeles County, population 9.8

million, in order to keep the narrative focused on the area surrounding the city.

Los Angeles, both city and county, comprises spread-out, polycentred areas, lacking a clearly defined core, as in New York, Chicago, and Toronto. The lack of dense urbanity makes the Los Angeles area a diffused metropolis, where highways knit together a checkerboard of neighbourhoods and towns of green lawns, detached houses, shopping strips, and malls. These are pleasant places of comfortable living, often enclosed by highways. The decentred structure of the area is also reflected in its small municipalities, except for the city of Los Angeles. Los Angeles County has eighty-eight municipalities, including the city, and acts as a regional government, providing police, fire, and other infrastructural services. Michael Dear and Nicholas Dahmann call Los Angeles a post-modern city-region in which the hinterland organizes the centre, development proceeds from outside-in, and fragmentation and decentredness are the primary development processes.[9] Recently, there have been attempts to develop the city's downtown and make it an art-cultural hub, with rapid transit linking it to the suburbs.

Los Angeles had Mexican roots, which made it a bicultural and binational city in its origin where settlers came to dominate. Unlike New York and Toronto, Los Angeles was not primarily an immigrants' town, though migrants came from the rest of the United States. The post-1965 wave of immigration has brought an exploding multi-ethnicity to the Los Angeles area. Its proximity to Mexico and connections with the Pacific Rim countries has put it at the confluence point of the two global channels of trade and migration, namely, the North-South Americas and the US-Pacific Asia. Mexicans and other Latin Americans have come to be the primary group of immigrants – some undocumented – but Chinese, Japanese, Koreans, and Filipinos also have a sizeable presence.

Los Angeles has become a majority-minority area, but its trajectory is to become a Latino metropolis, though Asians are also imprinting their cultures on the landscape and institutions, by both their growing population and economic entrepreneurship. Americans, Black and White, are now shrinking minorities.

Los Angeles County and the city are multicultural places in the same way as New York and Toronto are. The preceding section documents their similarities. Where multicultural Los Angeles differs from New York or Toronto lies in its urban structure, politics, and civic culture. Its small municipalities and autonomous communities have provided the scope for the local domination of one or another ethnic group, that is, geographic diversity from the bottom up. A number of towns that were passing English-only regulations in the 1980s – for example, Monterey Park, San Marino, and Walnut – are now steeped in Chinese or Latino cultures and languages.

Politics is the instrument for carving paths to incorporate ethno-racial groups. It is also the prime agent in organizing the "local context" of multiculturalism.[10] The Los Angeles area's political culture promoted the formation of informal political clubs of influential businessmen, ranchers, and real estate interests, which ran the local governments. This changed as the recent wave of immigrants started organizing for a share in the power.[11] By the 1980s, the power structure opened up. The traditional city leaders were ousted by ethno-racial voters in strategic alliance with Anglo-Jewish liberals. Locality after locality elected a new breed of leader, many of minority backgrounds.

This decentralized and fragmented (post-modern) process of injecting pluralism into institutions has resulted in producing the territorial quilt of subcultures and multicultural policies. One jurisdiction provides multilingual (translation and interpretation) access to services, while a neighbouring locality enacts laws for English only as the official language. The same institution may have different forms in different localities. Yet fundamental civil rights bridge these differences. Joel Kotkin, commenting on the fluidity of Los Angeles's cultural diversity, observes: "Los Angeles is constantly reinventing itself, combining and recombining people and neighborhoods from the ground up."[12] Like New York's, Los Angeles's multiculturalism is also led by private initiatives. It is driven by individual liberty and civil rights and not so much by public policy.

New York is the US media capital and Los Angeles is the global centre of films. Together, they produce the popular culture and will influence the forms that multiculturalism will take in the United States.[13]

The Differentiation and Integration of Diversity

The preceding description of the three cities suggests a paradigm of a multicultural city. Each city has been deeply invested with ethnic and racial diversity bolstered by civil rights. Pluralism has infused their urban structures and institutions. A resident or visitor to any of the three may be hard pressed to differentiate one city from the other on the basis of their everyday life.

A wide variety of dress, languages, accents, foods, music, and faces are jostling for one's attention in each city. An aura of normalcy hangs over such differences, with no hesitation and little self-consciousness about being different. The expectation that a uniformity of customs and behaviour will prevail has practically lost its hold in these cities. As discussed above, the similarities of multicultural cities are evident in the emergence of ethnic neighbourhoods, the formation of economic and commercial niches, the transformation of public places with the sounds, symbols, and smells of distant lands, the existence of culturally appropriate community services, and the entrenchment of a pluralistic ethos in the city's institutions, public policies, as well as market practices. These patterns of infusion of multiculturalism in the urban structure and community institutions are sustained by the social ethics of recognition and accommodation of cultural and racial differences.

Local factors and national creeds bring about differences in the forms of multiculturalism among cities. At the same time, within cities, these factors help integrate diverse forms. Toronto, New York, and Los Angeles are similar in multicultural patterns, but they differ in the modes of constructing and representing race and ethnicity. The forms and composition of their institutions may differ though functions may be similar. A few examples will illustrate these points.

All three cities have Chinese-controlled sectors and sub-economies, whose locus now is in the suburbs. As economic organizations, they have similar functions, namely, to organize a protected labour market for Chinese immigrants and entrepreneurs and specialize in producing goods and services for both the ethnic and mainstream markets. Yet they differ in their forms, composition, and local economic roles. Los Angeles's Chinese economy is closely linked with trade and manufacturing in Far Eastern countries, a role that is well suited to the city's emerging focus on the Pacific trade. It is dominated by Taiwanese investors, real estate developers, and professionals. New York's Chinese economy is based on the tourist trade, garment production, and the ethnic consumer market. It draws working-class immigrants from mainland China and nurtures professionals from the second generation. The city's generous social services have provided a cushion for struggling immigrants from the mainland and other Southeast Asian countries. Toronto's Chinese economy is first inward-looking, serving the ethnic market, but second is integrated into the economic base of the region through its niches in computer industries, tourism, and the fashion trade and, increasingly, in real estate. Hong Kong's influence is prominent in this economy. The three Chinese economies appear similar, but they differ in activities, nationalities, and forms. These differences are essentially the result of the distinct economy of each city.

Another example of local and national factors structuring multicultural institutions is that of the social construction of race and ethnicity and the politics of incorporating immigrants.[14] The term "White" in New York refers to Euro-Americans of different ethnicities. Blacks are African Americans, but the term popularly extends to West Indians. Spanish-speaking Puerto Ricans and Dominicans, though of dark skin, are identified as Hispanic. The Los Angeles area's ethno-racial construction is based on Anglos being the designation for Whites, who are differentiated from Chicanos and Mexicans, who become Latinos, though many of them are the successors of the Mexicans who lived in California before the conquest. Chinese, Koreans, Japanese, and Indians are subsumed into the category of Asians, with a "model minority" sobriquet. These differing narratives of race and ethnicity have a bearing on employment opportunities, social networks, and political organization.

Ethnicity plays a larger role than race in Toronto's discourse, though incidents of discrimination in jobs and public services lump non-Whites together. Whites are divided into English, Jews, Italians, East Europeans, and Portuguese. Hong Kong nationals have defined the ethnic identity of the Chinese community, though with increasing number of mainland Chinese in the city, linguistic differences among Chinese are cropping up. Indians, Pakistanis, Tamils, and others become South Asians. Official multiculturalism relies on these ethnic categories for both political discourse and economic opportunities. What can be seen in this description is that the same group has different identities in the ethno-racial pantheons of the three cities. These differences are reflected in the social, economic, and geographic patterns of the three cities.

Lastly, the differences of American and Canadian political cultures and national creeds filter down into the multicultural patterns and institutions of the three cities. The American discourse of individual liberty and private initiative promotes decentralized, flexible, and bottom-up forms of multicultural institutions. New York's cosmopolitanism lays the ground for a civic culture of assertive citizenship, which seeps into cultural and racial differences. Los Angeles's bi-nationalism and bilingualism spawned the politics of negotiated accommodations, which goes through cycles of ethno-racial confrontations and bargaining. Toronto is a piece of the Canadian social fabric. Its multiculturalism is initiated by communities' advocacy, but co-opted by public policy in a top-down institutionalism.

Multiculturalism is seldom talked about in the popular culture of the American cities, though ethno-racial issues are always in the media. In the Toronto area, multiculturalism is the idiom of everyday talk, with even banks and barbers advertising their multilingual credentials.

The City as a Common Ground: Place Matters

Multiculturalism is grafted on to an existing city that has a long-established territorial organization, physical infrastructure, and legal and socio-economic institutions. These structures and institutions contain and condition the cultural diversity of a city. They are elements of the common ground that is a reciprocal part of cultural diversity in multiculturalism.

The concept of common ground refers to the shared space, national and local laws, official language(s), infrastructure, governance system, economic organization, values, norms, and ideologies of national scope, public policies, market institutions, everyday etiquette, and moral codes – all that binds members of a society together. It translates the shared understanding and common interests into the routines and ethics of living together as a functioning nation and society.

Multiculturalism has built into it both cultural diversity and common ground. Living together requires not only a shared space and undivided infrastructure, but also the laws, norms, and values of a common civic culture, including some form of lingua franca to communicate and work through webs of mutual interdependencies and daily encounters.

The role of a city as a common ground is evident in the three cities. Whether it is Toronto, New York, or Los Angeles, its land-use pattern, urban form, housing quality, utilities and services, and architecture have a determining influence on the quality of life of everyone. Their city plans are conceived in terms of policies directed at organizing common factors such as densities, land uses, roads, streets and transportation networks, urban design, provision of services and amenities, and management of waterfronts and the environment. They aim to enhance the health, convenience, and welfare of people and realize a sustainable environment for all with equity, regardless of race and ethnicity. This is the ideal.

Also, the cities' iconic symbols, such as New York's Empire State Building and new One World Trade Center, Toronto's CN Tower, and Los Angeles's Hollywood sign, and now rising downtown towers, are sources of civic identity for all residents. The Metropolitan and Modern Art museums of New York, Toronto's Royal Ontario Museum, and Los Angeles's Getty Center and the Museum of Contemporary Art are similarly citizens' institutions, though they accommodate multiculturalism in their programs. These examples point out that cities are tied together by facilities, symbols, sports, art and aesthetics, over and above cultural differences.

Aside from national constitutions and laws, a host of city codes, by-laws, and regulations (such as those for traffic control, waste collection, zoning, and building), and mores of social activities (for example, those for getting in and out of buses and subways) define the norms and morals of public behaviour for all residents. These examples illustrate how a city conditions and bridges cultural differences through shared expectations.

Of course, a common ground evolves as a part of the process of continuing economic and social change. It is not carved in stone. Accommodations add to the dynamics of change by injecting new norms and values into the mainstream, prompting the process of forging a new synthesis. Cumulatively, the harmonization of accommodations leads to the restructuring of a common ground. It is an ongoing process, partially deliberated and partially happenstance, generating both dissent and consent. This is how multiculturalism becomes not only a process of implanting islands of cultural diversity in a sea of mainstream uniformity, but also an instrument for transforming urban institutions. The restructuring of a common ground proceeds in parallel with the incorporation of ethno-racial diversity.

There is another side to a common ground. It seeps into community cultures and prompts their cultural change. Witness, for example, the diffusion of American and Canadian music, fashions, and values, such as those of individualism and freedom, among the second generation of immigrants. The Chinese's own stereotype of their second-generation youth captures this change; that they are bananas, yellow on the outside and white inside.

Mainstream Society, Common Ground, and the Limits of Multiculturalism

Understanding the role of the common ground brings up the question of mainstream society's contributions to multiculturalism. The mainstream society is the source of energy and resources on which ethnic cultures and racial communities thrive. Its economy provides the opportunities for ethnic specializations and the formation of economic niches. Its technology, science, health and welfare services, law and order, and governance and educational institutions produce the good life that ethno-racial communities lay claim to and partake of. In the same vein, the mainstream society's liberal ideology, civil and human rights legislation, political institutions, and social organization create the "space" for the exercise of the entitlements of citizenship and the carving out of ethnic and racial identities. All in all, mainstream society is the medium within which ethnic communities grow and multiculturalism is cultivated.

Mainstream society's role as the medium for the growth of ethno-racial communities and expression of their identities has not been recognized in the literature on multiculturalism. The left-liberal strain of the literature views the mainstream society as an obstacle in realizing multiculturalism's goals of diversity and equality. The narratives of discrimination and racism essentially blame it for these conditions.

Mainstream society both nurtures and defines ethno-racial differences. It defines the scope and limits of multiculturalism. Yet it has a foundational role in the multiculturalism that is practised within the shared space of a city. The territorial multiculturalism of regional cultures in a nation is of a different order, such as those of Quebec and Newfoundland in Canada, the North and South in the United States, or Scotland in the United Kingdom.

Democracy, equality, individualism, freedom of expression, the rule of law, tolerance, civility, and other basic values, norms, and morals of the mainstream society are the cultural universals of a multicultural community and society. They define the scope and limits of cultural differences. Ethno-cultural practices that violate these basic values and morals are discouraged. The cultural differences that fall within the basic values are recognized and accommodated. This is the meaning of reasonableness in the policies of reasonable accommodation. The mainstream society also mediates conflicts of values and practices among ethnic cultures and communities. It is necessary for the functioning of interculturalism, a term coming into use in preference to multiculturalism. On all these scores, the mainstream society is an essential part of the common ground necessary for the functioning of a multicultural community. It deserves to be appreciated.

Multiculturalism Revisited

The evidence from Toronto, New York, and Los Angeles bears out the validity of the model of multiculturalism developed in chapter 2. The preceding sections of this chapter have summarized the evidence of the three cities having developed by balancing both dimensions of multiculturalism. Yet this evidence also reflects upon the theories and concepts of multiculturalism, which are discussed in chapter 2 and the appendix.

The discourse about multiculturalism essentially revolves around two themes, namely, (1) the entitlement of individuals and communities to the recognition and expression of their identities and accommodation of their cultural, racial, and religious differences and (2) the unity of the constitution, laws, values, ethics, economic institutions, and common language(s) required for living together in a shared space, society, and state. In other words, multiculturalism is a matter of balancing diversity rights and the imperatives of societal integration.

The theorizing about different forms of multiculturalism, that is, liberal, conservative, and communitarian, proceeds by postulating a varying "mix" of the entitlements of diversity, on the one hand, and social cohesion, on the other. These theories are primarily conceived in terms of political rights and institutions. But multiculturalism is also about "culture" and social organization, matters to which these theories are indifferent.

The empirical observations of the multiculturalism of the three cities point out the primacy of the cultural, social, and economic changes induced by the infusion of multiculturalism. Backed by human and civil rights, multiculturalism is also about beliefs and behaviours, common rules and shared expectations, reasonable accommodations and functional imperatives, expressions of identities and civic spirit. The preceding chapters show that the multiculturalism of the three cities is as much the outcome of the accommodations of ethno-racial differences as of the bridging influences of space, infrastructure, laws, values, and policies. The evidence bears out Will Kymlicka's argument that in multiculturalism the incorporation of differences goes along with a reiteration of national identity and cultivation of loyalties through education, citizenship, language, media, and history.[15] In the case of cities, space and collective goods are also means of integration. All in all, diversity rights and the common ground evolve in tandem as the two sides of multiculturalism. The historic mainstream society is the medium into which multicultural ethos is infused. It is an active agent in both accommodating diversity and defining its scope and limits.

The findings of this book inductively affirm the five aspects of multiculturalism postulated in chapter 2, namely, (1) the right to exercise cultural and religious beliefs and behaviours in the mostly private domain individually and as communities; (2) a unitary public domain that is inclusive of different identities and non-discriminatory; (3) a common ground overlapping with the public domain but going beyond it to include entitlements of citizenship and the shared norms, values, and everyday practices required by the interdependencies of living together, particularly in cities; (4) the two-sidedness of multiculturalism: diversity rights and the common ground; and (5) a pluralistic civic culture and social order that is a continuing process of balancing differences and promoting integration.

Another theme that has emerged from this book is the limited scope of the cultures of multiculturalism. They are subcultures that do not affect many areas of social life, such as criminal and civil laws, human rights and universal values, economic institutions, technology, national language(s), land, and infrastructure. Inherent in the idea of subculture is the limits of its scope. This applies to cultures of multiculturalism. Many controversies of multiculturalism arise from a failure to realize the subcultural nature of ethno-racial cultures in a society.

Another common assumption in the discourse about multiculturalism is the essentialism of ethno-racial cultures and lifestyles. All cultures are dynamic, changing with exposure to new ways of life, values, and technologies. The dynamics of cultural change has to be given attention in discussions of multiculturalism.

Finally, the discourse on multiculturalism does not give any weight to the spatial dimension of social and economic organizations. Space is a strong variable in multiculturalism. Sharing the defined space of a neighbourhood and city as well as its associated public utilities and services results in interdependencies of mutual behaviours. This sharing requires common practices cutting across "cultural" differences. Observing multiculturalism in the concrete context of the three cities has highlighted the unifying role of space and its associated civic culture.

At this juncture in history, multiculturalism is viewed with some scepticism, particularly in Europe, where it was embraced enthusiastically initially but has been hit by a political backlash. In recent times, leaders of Britain, France, and Germany have called for abandoning multiculturalism, because of its presumed impeding of immigrants' integration.[16] Race riots, acts of terrorism, and economic recession have combined with the rise of right-wing parties to sow distrust of immigrants in general and Muslims in particular. Thus, the politics of immigration and national security gets entangled with attitudes towards multiculturalism.

Problems of immigrants' acculturation and integration are not to be laid entirely at the door of multiculturalism. The settlement and integration of immigrants is a distinct process with its own dynamics and politics. Multiculturalism applies as much to the native-born as to immigrants, who become citizens and turn into ethnics of particular identities in subsequent generations.

Western societies including Canada and the United States, are not reproducing their populations. They are now dependent on immigration to sustain their economic growth and social stability. Immigrants come with their identities, memories, beliefs, behaviours, and customs, which do not melt away on crossing borders. They come on the promise of equal rights, including the right to their identities. Also, transnationalism is an increasing trend, bringing about dual nationalities. Pluralism is the only model that applies to a society drawn from diverse populations and global connections. In cities of majority-minority populations, multiculturalism is the only feasible urban culture.

Multiculturalism is always a work in progress. One need be neither romantic nor jaded about its effects. It does not by itself eliminate racism, ethnic conflicts, and ethno-political competition. It does not result in banishing controversies about matters such as the hijab, prayer rooms in colleges, discrimination in employment, or attitudes towards women and gays, for example. Yet as a normative model, it offers a path to manage such conflicts and promote social harmony by demarcating rights and responsibilities. Again, a robust common ground is the key to this promise.

At its core, multiculturalism is about equity in cultural rights and the recognition of differences in lifestyles and beliefs. It is not a sufficient condition for economic equity and social justice. Yet it promises to reduce inequities on the basis of cultural and racial identities.

Multiculturalism exists with or without official policies. The United States is not an officially multicultural country, but practically it is as multicultural as Canada with its official multiculturalism. It has a long tradition of cultural pluralism arising from its ethos of liberty. Ethnocultural diversity has been a part of US city life since the arrival of South and East European immigrants in the early twentieth century. Even in

Europe, as countries disavow multiculturalism, everyday life in cities remains multicultural. Civil rights and market processes sustain cultural diversity. Jan Rath observes about Europe, "Despite complaints about immigration and diversity, and despite integrationist and assimilationist discourses, 'multiculturalism' by stealth is de rigueur."[17] Doug Saunders heartily agrees with this observation.[18]

The City of the 21st Century: Multicultural

To sum up the findings of this book, a multicultural city can be defined to have six characteristics: (1) an ethno-racially diverse population; (2) a regime of civil/human rights that enacts cultural and religious freedoms, recognizing and promoting diverse identities and subcultures in the private domain of home and community; (3) an institutionalized common ground of basic societal values, ideologies, constitution and laws, norms, moral codes, and official language(s) as well as shared space and services; (4) reasonable accommodations of cultural differences that make public institutions pluralistic and inclusive; (5) a harmonization of differences and reconstruction of mainstream institutions to promote shared citizenship, a sense of belonging, and a common civic culture; and (6) a changing subculture and common ground through diffusion of influences from one to the other. This city meets the needs of the twenty-first century.

The city in the twenty-first century is going to be shaped by the movement of populations, international networks, and a rapid circulation of ideas, finances, technologies, and products. In North America, demography is the driving force in a major realignment of social and cultural relations. In major North American cities historically dominant groups are not in the majority anymore. The changing demography has eroded the myth of a homogeneous culture. Globalization is further breaking down cultural insularity. The national culture is going to be an evolving mixture of cultures. A variety of ways of life and beliefs are going to be recognized within a framework of universal values and national ideologies. All these trends come together in cities. They cannot but be multicultural.

The multicultural city represents an implicit social contract that might be pursued more explicitly. Immigrants and ethno-racial minorities have an entitlement to the recognition of their identities and equality of social rights. In return, they have the responsibility to embrace the common ground and contribute to collective goals and values. Dowell

Myers, drawing on Beth Rubin's definition of the social contract, points out that people of unequal resources pursuing self-interest may work towards shared social understanding through structured cooperation.[19] He goes on to posit a social contract between aging boomers and young immigrants, whereby the boomers now support educational and social services for immigrants, who in turn will undertake to fund the former's old age security, both through taxes.[20] Myers's idea of a social contract can be applied to cultural rights, the integration of immigrants, and building of a robust common ground. This is the ideal of a multicultural city. It beckons North Americans in the twenty-first century.

Theoretical Discourse on Multiculturalism

Forms of Multiculturalism

This appendix complements chapter 2. It discusses theories and arguments about the scope of multiculturalism as a political ideology; the discussion that could have distracted from the main focus of chapter 2, namely, developing a workable model of multiculturalism to guide empirical observations in subsequent chapters. That discussion has been brought into this appendix for those readers interested in examining the current debates in multicultural theories.

Regarding the adoption of the multicultural ethos by a nation state, Kymlicka postulates three principles, namely: (1) repudiation of the idea of national homogeneity and the privileging of one cultural group, (2) repudiation of the policies of assimilation and of enabling minorities to access state institutions without compromising their ethno-cultural identity, and (3) acknowledgment of the historic injustices to minorities and their remediation.[1] The enactment of these principles realigns the public domain. The extent to which these principles are applied determines the breadth and depth of ethno-cultural diversity in a society. Yet these are political decisions. That is why much of the theorizing on multiculturalism is conceived in terms of political rights and institutions.

The essence of multiculturalism is cultural pluralism as a matter of public right, which takes many forms. The specific form multiculturalism takes in a country depends on its ethno-racial mix, constitutional structure, ideology, and political culture. These factors affect the enactment of Kymlicka's three principles. They also spawn the politics of rights and identities, which in turn determine the form of multiculturalism in a particular country. To begin with, there are two broad forms of multiculturalism: (1) official or legislated and (2) de facto or lived.

Canada is an officially proclaimed multicultural society. The United States has a de facto or lived multiculturalism nurtured by the Civil Rights Act (1964), anti-racism policies, and equality legislations. Both are nations of immigrants. Canada's multiculturalism arose out of its long political struggle to accommodate the national interests of French Canadians and aboriginals. It has been founded not on the idea of national homogeneity but on that of a civic nation tied together by citizenship and common values. Its multinational and bilingual heritage laid the ground for Canada to recognize immigrants' cultures and identities in a legislated multicultural policy (1971) and, finally, the Multiculturalism Act (1988).

In the United States, the Civil Rights Act (1964) and the assertion of Blacks' cultural pride challenged the notion of a uniform American culture in the 1960s. These movements lent a renewed legitimacy to the ethnic identities that have always been a defining element of the American social organization. American multiculturalism is a byproduct of its history, social organization, and civil rights. It is a lived multiculturalism. Sociologists have almost invariably recognized the cultural pluralism of the United States.[2] These differences in origins give a distinct colour to the multiculturalism of each country. Canadian multiculturalism is defined by its policies of recognizing ethno-cultural and linguistic diversity, while in the United States race/ethnic relations, education, and politics are the main arenas of cultural pluralism.

Politically, multicultural rights are packaged in different forms. Angie Fleras postulates three models of multiculturalism.[3] (1) Conservative multiculturalism is libertarian in spirit. It envisages no special treatment for anyone, but everyone is entitled to equal citizenship without any bar of colour or culture. It recognizes individuals' differences, but is indifferent to community identities. (2) Liberal multiculturalism is based on principles of diversity and equality. Minorities have rights to their cultural identities and an entitlement to equality with all their differences. "People are treated similarly by taking differences into account when necessary to ensure equality."[4] (3) Plural multiculturalism envisages collective group rights implemented through autonomous institutions and separate communities. In this model diversity flourishes in separate institutions of equal standing.[5] It can also be characterized as communitarian multiculturalism, as it privileges communities over individuals. In it ethnicity is destiny.

In the modern democratic states of Canada and the United States, with their regimes of civil and human rights and political equality, liberal multiculturalism is the reigning form. In practical terms, liberal multiculturalism is implemented by policies such as (1) legislative affirmation of multiculturalism and minority rights; (2) a multicultural curriculum in schools; (3) diverse dress and holiday codes; (4) the funding of ethnic and cultural activities; (5) the funding of educational (language) and interpretation services; (6) affirmative action for minorities and immigrants; and (7) accommodation of cultural and religious differences in the provision of services.[6] These policies are followed to varying degrees in both Canada and the United States.

As Howard Duncan has aptly put it: "If you do not have liberalism you are not going to have multiculturalism."[7] Will Kymlicka, the leading theorist and advocate of minority rights also places multiculturalism squarely within liberal ideologies.[8] He argues that in Canada "[the] reform of diversity policies is simply one more example of [this] liberalization ... enacted by the same liberal reformist coalitions."[9]

There are some variations in the modes of accommodating cultural diversity that give different weights to the integrative institutions of the public domain versus the cultures of the private domain. Interculturalism is one of those notions. It is subtly distinct from multiculturalism in that it envisages "creating a single [but] diverse public not multiple publics."[10] The province of Quebec in Canada defines its policy as interculturalism, in distinction from federal multiculturalism, purportedly laying greater weight on the unifying public domain to "reconcile ethno-cultural diversity with the continuity of the French-speaking core and preservation of the social link."[11] It emphasizes a strong common ground of shared values and language. The goal in various formulations of multiculturalism is to balance the demands of national integration with diversity rights in a framework of equality.

Critiques of Multiculturalism

Multiculturalism as an ideology or policy is a contested idea. It is opposed on many grounds, conceptual, ideological, and political. Conceptually, there is a fundamental tension between the recognition and expression of cultural diversity, on the one hand, and the pull to integrate multicultural communities into the mainstream and instil in them a sense of belonging, on the other.[12] Ideologically, multiculturalism sits well with the left-liberal values of social justice, individual rights, and freedoms. Yet many of those espousing the rights of minorities and the disadvantaged find multiculturalism to not be fulfilling its promise of equality and fairness. For them, it ignores the racism that has long afflicted minorities by shifting the focus onto culture instead of race and class. Another criticism from this perspective is that multiculturalism ghettoizes minorities and freezes them in primordial identities, thereby reinforcing conservative mores in community cultures.[13] In the same vein, multiculturalism is said to be promoting the exoticism of song and dance rather than liberal values and human rights. Some argue for a post-multiculturalism and post-ethnicity that promotes human rights and ethno-racial equality without herding individuals into ethnic enclosures.[14] The underlying theme of these criticisms is that multiculturalism goes only part way towards achieving equality of diversity.

From the conservative perspective, multiculturalism emphasizes a group or community over the individual and distracts from liberty. Yet the strongest critiques of multiculturalism from this perspective are couched in terms of its social divisiveness and threat to national cohesion and security. It takes away from social integration and encourages immigrants to resist assimilation.[15] Post-9/11 concerns about terrorism in Western societies have given a new urgency to the integration of minorities and immigrants and infusing them with national identity. The acts of terrorism in London, Copenhagen, and Madrid and ethnic riots in Paris suburbs have driven Britain, the Netherlands, and Denmark to retreat from multiculturalism. In Canada, frequent voices are raised to express the sentiment "Enough of multiculturalism, bring on the melting pot."[16] In the United States, anti-immigration sentiments, the English-as-the-national-language movement, and libertarianism are raising questions about diversity rights, as has the national concern with terrorism.

Politically, multiculturalism is mixed up with the problems of immigration. The two are overlapping but distinct phenomena. The former is about the pluralistic structure of a society in which cultural rights are recognized and accommodated. The latter is the process of bringing immigrants into a country and organizing their settlement, acculturation, and integration. Multiculturalism applies to both immigrants and native-born ethnics.

Periodically, anti-immigrant sentiments are ignited by right-wing political groups, usually in periods of an economic down cycle, both in European countries and in parts of the United States and Canada. Thus, the politics of immigration sweeps up multiculturalism into its fold. When the British prime minister or German chancellor proclaims the failure of multiculturalism, they are scoring political points with their conservative constituencies.[17] Yet simultaneously they promote immigration to counter their aging and shrinking populations and fill the demand for labour in their economies. Such are the contradictions of the politics of immigration, which are projected onto multiculturalism.

These swings of political sentiments reverberate in the multiculturalism discourse. It has swung from celebrating minority cultures in the 1980s, to emphasizing racism and inequality in the 1990s, to focusing on the need to promote social integration and build cohesive societies in the 2000s. Regardless of where the pendulum of public discourse rests at any one moment, multiculturalism will continue to be a lived reality. With large-scale and continual immigration, ethnic and religious diversity is being reinforced. Even if a country, such as the Netherlands or Britain, were to rescind its official multiculturalism, its ethnic differences will continue to thrive. Ethnic communities' civil rights to live their private life by their values and norms, enjoy freedom of religion, and have equal access to the market and state will sustain their diversity. They can neither be denied their identities, nor have their liberties taken away, without the abandonment of a society's fundamental rights. Unless there is a wholesale retreat from liberal democracy and human rights – something approximating totalitarianism – cultural diversity will remain a defining quality of North American societies in the twenty-first century. It will have to be accommodated in national institutions. Most of the arguments against multiculturalism arise from a one-sided view of diversity. Either diversity rights are highlighted or social integration is made to be the touchstone of a society. Yet the two sides are intertwined.

One point overlooked in the discussions of multiculturalism is the dynamic nature of cultures and subcultures. They are continually evolving and changing by both internal innovation and borrowing from others. The "cultures" of multiculturalism are not self-contained and closed. They are not only limited in scope but are also continually absorbing beliefs and practices from one another and the mainstream. One interesting idea about cultures' fluidity is offered by Arjun Appadurai: that culture is not just what has been followed in the past but also includes norms of future behaviour, which are influenced by aspirations about the future.[18] Hybridity and change are the defining conditions of cultures, both of the historic majority and of ethnic minorities. Kian Tajbakhsh maintains that urban identities are "never re-ducible to one thing," they are mixed up.[19] Bhikhu Parekh, an eminent British authority on multiculturalism and the chair of the Commission on the Future of Multi-Ethnic Britain (2000), says it well: "Cultures are not museum pieces but living, changing, thriving systems, and cultural diversity is a moving feast."[20]

Notes

1. Cultures and the City

1 The competition on the BBC international service was held on 24 August 2007.
2 Leonie Sandercock, *Cosmopolis II Mongrel Cities* (London: Continuum, 2003), 4.
3 Nathan Glazer, "Assimilation Today: Is One Identity Enough?" in *Reinventing the Melting Pot*, ed. Tamar Jacoby (New York: Basic Books, 2004), 727.
4 Nancy Foner, "How Exceptional Is New York? Migration and Multiculturalism in the Empire City," *Ethnic and Racial Studies* 30, no. 6 (2007), 999–1023.
5 The US will become a majority-minority country due to immigration between 2042 and 2050. Angela Blackwell, Stewart Kwoh, and Manuel Pastor, *Uncommon Common Ground* (New York: W.W. Norton, 2010), 29.
6 Roger Waldinger and Mehdi Bozorgmehr, "The Making of a Multicultural Metropolis," in *Ethnic Los Angeles*, ed. R. Waldinger and M. Bozorgmehr (New York: Russell Sage Foundation, 1996), 3–38. .
7 Will Kymlicka, *Multicultural Odysseys* (New York: Oxford University Press, 2007), 61.
8 Ash Amin, "Ethnicity and the Multicultural City: Living with Diversity," *Environment and Planning A* 34 (2002), 974.
9 Most of the authors who have written about multicultural cities proceed on the assumption that multi-ethnicity and its associated multiplicity of cultures are enough as indicators of multicultural places. Nancy Foner in examining the diversity in New York, Ash Amin talking about ethnicity in the multicultural city, or Roger Waldinger and Mehdi Bozogmehr

reviewing the making of Los Angeles as a multicultural metropolis do not define the parameters of the multicultural city.

10 J. Frow and M. Morris, eds, *Australian Cultural Studies: A Reader* (St Leonards, NSW: Allen and Unwin, 1993), viii.

11 UNESCO, "Universal Declaration of Cultural Diversity," 21 February 2002.

12 Definition derived from Claude Fisher, "The Subcultural Theory of Urbanism: A Twentieth-Year Assessment," *American Journal of Sociology* 101, no. 3 (1995), 544.

13 Sharon Zukin, *The Cultures of Cities* (Cambridge: Blackwell Publishers, 1995), 1.

14 Iain Mclean and Alistair McMillan, *Oxford Concise Dictionary of Politics*, 2nd ed. (New York: Oxford University Press, 2003), 77.

15 Zukin postulates the concept of public culture, defining it as modes of interacting in streets, parks, and shops and the rights emerging from such encounters. This concept overlaps with civic culture, though the latter includes formal rules and rights of participating and benefiting from governmental institutions. *The Culture of Cities*, 11.

16 Claude S. Ficher, "The Subcultural Theory of Urbanism: A Twentieth-Century Assessment," *American Journal of Sociology* 101, no. 3 (1995), 568.

17 Barry Wellman and Barry Leighton, "Networks, Neighborhoods, and Communities: Approaches to the Study of the Community Question," *Urban Affairs Quarterly* 14, no. 3 (1979), 363–90.

18 Kymlicka, *Multicultural Odysseys*, 44.

19 David Brooks, *Bobos in Paradise: The New Upper Class and How They Got There* (New York: Simon and Schuster, 2001).

20 Richard Alba and Victor Nee, *Remaking the American Mainstream* (Cambridge, MA: Harvard University Press, 2003), 11. Waldinger and Bozorgmehr, "The Making of a Multicultural Metropolis," 30.

21 Adapted from Wsevolod Isajiw, *Definitions of Ethnicity*, Occasional Papers in Ethnic and Immigration Studies (Toronto: Multicultural History Society of Ontario, 1979), 25.

22 K. Anthony Appiah, "Identity, Authenticity, Survival," in *Multiculturalism*, ed. Charles Taylor (Princeton: Princeton University Press, 1994), 160.

23 Amartya Sen, *Identity and Violence* (New York: W.W. Norton, 2006), 12.

24 That is, the right to participate in city life and governance.

2. Diversity Rights and the Common Ground

1 Paraphrased from Peter Kivisto, "We Really Are All Multiculturalists Now," *Sociological Quarterly* 5, no. 3.1 (2012), 4.

2 John Rex, "The Concept of a Multicultural Society," in *The Ethnicity Reader*, ed. Montserret Guibernam and John Rex (Cambridge: Polity Press, 1997), 207.

3 John Rex, *Ethnic Minorities in the Modern Nation State* (Houndmills: Macmillan Press, 1996), 16.

4 Initially, in 1971, Canada adopted the Multiculturalism Policy, which was incorporated in the Canadian Multiculturalism Act of 21 July 1988.

5 Will Kymlicka, *Multicultural Odysseys* (New York: Oxford University Press, 2007), 83.

6 The Parekh Report, *The Future of Multiethnic Britain* (London: Profile Books, 2000), 43.

7 Rex, "The Concept of a Multicultural Society," 19.

8 Kivisto, "We Really Are All Multiculturalists Now," 4.

9 Angie Fleras, *The Politics of Multiculturalism* (New York: Palgrave, 2009), 8–11.

10 Kymlicka, *Multicultural Odysseys*, 61.

11 Charles Taylor, *Multiculturalism* (Princeton: Princeton University Press), 28, 38.

12 Kivisto, "We Really Are All Multiculturalists Now," 5.

13 Tariq Madood, *Multiculturalism* (Cambridge: Polity Press, 2007), 2, 44..

14 Kymlicka, *Multicultural Odysseys*, 83.

15 Ibid., 83.

16 Fleras, *The Politics of Multiculturalism*, viii.

17 Rex, "The Concept of a Multicultural Society," 18.

18 Will Kymlicka, *Finding Our Way* (Toronto: Oxford University Press, 1998), 58.

19 Keith Banting, Thomas Courchene, and Leslie Seidle, "Introduction," in *Belonging? Diversity, Recognition and Shared Citizenship in Canada*, ed. K. Banting, T.J. Courchene, and F.L. Seidle (Montreal: Institute of Public Policy, 2007), 11–12.

20 The Parekh Report, *The Future of Multi-Ethnic Britain*, 46.

21 Horace Kallen, *Culture and Democracy in the United States: Studies in the Group Psychology of the American People* (Salem, NH: Ayer Company, 1924).

22 Nathan Glazer and Daniel Moynihan, *Beyond the Melting Pot* (Cambridge, MA: MIT Press, 1963), 310. 313–15.

23 Nathan Glazer, "Beyond the Melting Pot: Thirty Years Later," in *Lectures and Papers in Ethnicity, no. 5,* July 1991 (Department of Sociology, University of Toronto, 1991), 11.

24 Nathan Glazer, *We Are All Multiculturalists Now* (Cambridge, MA: Harvard University Press,1997), 159.

25 Kivisto, "We Really Are All Multiculturalists Now," 20.

26 Fleras, *The Politics of Multiculturalism*,110.

27 Mohammad Qadeer, "The Charter and Multiculturalism," *Policy Options* 28, no. 2 (2007), 90–1.

28 Parekh Report, *The Future of Multiethnic Britain*, 44.
29 David Hollinger, *Postethnic America: Beyond Multiculturalism* (New York: Basic Books, 2000), 96–7.
30 Ash Amin and Nigel Thrift, *Cities: Reimaging the Urban* (Cambridge: Polity, 2002), 145–8.
31 Robert D. Putnam, *Bowling Alone* (New York: Simon and Schuster, 2000), 22.
32 Ashtosh Varshney, *Ethnic Conflict and Civic Life* (New Haven: Yale University Press, 2002), 281–2.
33 *New York Times*, 27 May 2009, http://www.nytimes.com/glogin?URI=http://www.nytimes.com/2009/05/28/style/28hugs.html&OQ=_rQ3D0&OP=4491077dQ2FrQ2AQ251rX31rFFFrQ511Q26Q2BrjQ2AUxyQ2AQ2A1Q20rQ20Q7DQ7DhrQ7DbrQ209rx13Q2BirQ209Q51(_xYQ511Q26Q2B.
34 Richard Alba and Victor Nee, *Remaking the American Mainstream* (Cambridge, MA: Harvard University Press, 2003), 64.
35 Nathan Glazer, "Assimilation Today: Is One Identity Enough?" in *Reinventing the Melting Pot*, ed. Tamar Jacoby (New York: Basic Books, 2004), 73.
36 Herbert Gans, "The American Kaleidoscope, Then and Now," in *Reinventing the Melting Pot*, ed. Jacoby, 44.
37 Aljendro Portes and Ruben G. Rumbautm, *A Portrait of Immigrant America*, 3rd ed. (Berkeley: University of California Press, 2006), 271.
38 C.W. Watson, *Multiculturalism* (Buckingham: Open University Press, 2000), 3.
39 Paul Anisef and Michael Lanphier, "Introduction: Immigration and Accommodation of Diversity," in *The World in a City*, ed. P. Ansief and M. Lanphier (Toronto: University of Toronto Press, 2003), 6.
40 James Frideres, "Creating an Inclusive Society: Promoting Social Integration in Canada," in *Immigration and Integration in Canada in the Twenty-First Century*, ed. John Biles, Meyer Burstein, and James Frideres (Montreal: McGill-Queen's University Press, 2008), 81.
41 Tariq Madood, "Multiculturalism, Ethnicity and Integration: Contemporary Challenges," *Canadian Diversity* 5, no. 1 (2006), 120.
42 Gerard Bouchard and Charles Taylor, *Building the Future. A Time for Reconciliation*, Report of the Commission de Consultation sur les Pratiques d'Accommodement Reliées aux Différences Culturelles, (Quebec, 2008), 19.
43 Yasmeen Abu-Laban and Baha Abu-Laban, "Reasonable Accommodation in a Global Village," *Policy Options* 26, no. 8 (2007), 30.
44 Julius Grey, "The Paradox of Reasonable Accommodation," *Policy Options* 26, no. 8 (2007), 34–5.

45 Peter Hall, *Cities in Civilization* (London: Weidenfeld & Nicolson, 1998), 6.
46 Jane Jacobs, *The Death and Life of Great American Cities* (New York: Vintage Books, 1992), 14.
47 Richard Florida, *The Flight of the Creative Class* (New York: Collins, 2005), 62.
48 William Shakespeare, *Coriolanus*, act 3, scene 1.
49 Janet Abu-Lughod, *Changing Cities: Urban Sociology* (New York: HarperCollins, 1991), 140.
50 James Holston and Arjun Appadurai, "Cities and Citizenship," *Public Culture* 8 (1996),188–9.
51 Ibid., 200.
52 Ash Amin, "The Good City," *Urban Studies* 43, nos. 5/6 (May 2006),1012.
53 Susan S, Fainstein, *The Just City* (Ithaca; Cornell University Press, 2010), 3.
54 Ibid., 43.
55 Leonie Sandercock, *Mongreal Cities* (London: Continuum, 2003), 87.
56 Henri Lefebvre, *Writings on Cities*, trans. Eleonore Kofman and Elizabeth Lebas (Malden, UK: Blackwell Publishing, 2006), 179, 194.
57 Mark Purcell, "Possible Worlds: Henri Lefebvre and the Right to the City," *Journal of Urban Affairs* 36, no. 1 (2013), 148–9.
58 David Harvey, "The Right to the City," *International Journal of Urban and Regional Research* 27, no. 4 (2003), 941.
59 See also the Appendix.

3. Making Multicultural Cities

1 Dowell Myers, *Immigrants and Boomers* (New York: Russell Sage Foundation, 2007), 55.
2 Statistics Canada, *The Daily*, 9 March 2010, http://www.statcan.gc.ca/daily-quotidien/100309/dq100309a-eng.htm.
3 Myers, *Immigrants and Boomers*, 57, figure 3.7.
4 David K. Foot, *Boom, Bust and Echo* (Toronto: Macfarlane Walter & Ross, 1996), 18.
5 Statistics Canada, Birth and Fertility Data 1989–2009, CANSIM Table 102-4502, Catalogue no. 84-214-X-PE.
6 William Frey, "Shift to a Majority-Minority Population in the U.S. Happening Faster than Expected," 19 June 2013, www.brookings.edu/blogs/experts/freyw.
7 Myers, *Immigrants and Boomers*,193–4.
8 Metropolitan Program at Brookings, *The State of Metropolitan America 2010*, www.brookings.edu/research/reports/2010/05/09-metro-america.

9 David Halle and Andrew Beveridge, "New York and Los Angeles: The Uncertain Future," in *New York and Los Angeles*, ed. David Halle and Andrew Beveridge (New York: Oxford University Press, 2013), 5, table1.1.

10 Ibid.

11 William H. Frey, "Diversity Explosion: The Cultural Generation Gap Mapped," November 19,2014 www.brookings.edu.

12 William H. Frey, "Melting Pot Cities and Suburbs: Racial and Ethnic Change in Metropolitan America in the 2000s," *Metropolitan Policy Program at Brookings*, http://www.brookings.edu?papers/2011/0504_census-ethnic-frey.Aspx,1.

13 Ibid.,1.

14 See Vivek Wadhwa, "The Face of Success," 12 January 2012, http://www.inc.com/vivek-wadhwa/how-the-indians-succeeded-in-silicon-valley.html.

15 Christine M. Matthews, *Foreign Science and Engineering Presence in U.S. Institutions and the Labor Force*, Congressional Research Service Report, 7–5700, 4–5.

16 Philippe Legrain, *Immigrants: Your Country Needs Them* (London: Abacus, 2007), 110.

17 Nancy Foner, *In a New Land* (New York: New York University Press, 2005), 131.

18 Thomas Muller, *Immigrants and the American City* (New York: New York University Press, 1993), 161–215. George J. Borjas, "The Impact of Immigrants on Employment Opportunities of Natives," in *Immigration Reader*, ed. David Jacobson (Malden, UK: Blackwell Publishers, 1998), 217–30. Thomas J. Espenshade, "U.S. Immigration and the New Welfare State," in *Immigration Reader*, ed. Jacobson, 231–50.

19 Legrain, *Immigrants*.

20 Doug Saunders, *Arrival City* (Toronto: Alfred A. Knopf Canada, 2010).

21 Muller, *Immigrants and the American City*, 219.

22 J.S.Woodsworth, *Strangers within Our Gates. Or: Coming Canadians* (1909) (Toronto: University of Toronto Press, 1972), 155.

23 Quoted in Jeffrey G. Reitz, *The Survival of Ethnic Groups* (Toronto: McGraw-Hill Ryerson, 1980), 7.

24 Samuel Huntington, *Who Are We? America's Great Debate* (New York: The Free Press, 2005).

25 Martin Collacot, *Submission to the Commission on Accommodation Practices Related to Cultural Differences* (Vancouver: Fraser Institute, 2007), http://www.Fraserinstitute.org/research_news/display.aspx/id=13433.

26 Janice G. Stein, "Searching for Equality," in *Uneasy Partners*, ed. Janice G. Stein, David R. Cameron, et al. (Waterloo: Wilfrid Laurier University Press, 2007), 17.

27 Michael Adams, *Unlikely Utopia* (Toronto: Viking, 2007), 10.

28 Joseph Berger, *The World in a City* (New York: Ballantine Books, 2007), 264.

29 E.g., Muller, *Immigrants and the American City*.

30 Kwame Appiah, "The Case for Contamination," *New York Times Magazine*, http://www.nytimes.com/2006/01/01/magazine/01cosmopolitan.html?.

31 Pico Iver, *The Global Soul* (New York: Viking Books, 2000).

32 Michael Valpy "Our Part-Time Home and Native Land," *Globe and Mail*, 28 June 2008, A13.

33 Valpy, "Our Part-Time Home."

34 Lima, "Living Here and There."

4. The Social Geography of Multicultural Cities

1 David F. Ley and Larry S. Bourne, "Introduction: The Social Context and Diversity of Urban Canada," in *The Changing Social Geography of Canadian Cities*, ed. L.S. Bourne and D.F. Ley (Montreal: McGill-Queen's University Press, 1993), 3.

2 Ernest W. Burgess, "The Growth of the City," in *The City*, ed. Robert Park et al. (Chicago: University of Chicago Press, 1967).

3 Homer Hoyt, *The Structure and Growth of Residential Neighbourhoods in American Cities* (Chicago: University of Chicago Press, 1939). Chauncy Harris and Edward Ullman, "The Nature of Cities," *Annals: American Academy of Political and Social Sciences* 242 (1945), 7–17.

4 Michael Dear and Nicholas Dahmann, "Urban Politics and the Los Angeles School of Urbanism," in *The City Revisited*, ed. Dennis R. Judd and Dick Simpson (Minneapolis: University of Minnesota Press, 2011), 68–9.

5 David Halle and Andrew A. Beveridge, "The Rise and Decline of the L.A. School and New York Schools," in *The City Revisited*, ed. Judd and Simpson, 139.

6 Bill Bishop, *The Big Sort: Why the Clustering of Like-Minded America Is Tearing Us Apart* (New York: Mariner Books, 2008).

7 Alejandro Portes and Robert Bach, *Latin Journey: Cuban Immigrants in the United States* (Berkeley: University of California Press, 1985).

8 *Toronto Star*, "High-Rise Ghettos," Life Section, 3 February 2001, M1–3.

9 Peter Marcuse, "Enclaves Yes, Ghetto No: Segregation and the State," in *Desegregating the City*, ed. David P. Varady (Albany: State University of New York Press, 2005), 16–17.

10 Ceri Peach, "The Ghetto and the Ethnic Enclave," in *Desegregating the City*, ed. Varady, 31–48. Peter Marcuse, ibid., 15–30.

11 Marcuse, ibid., 18.

12 *Missionary Outlook* (Toronto) 30, no. 12 (12 Dec. 1910), 267.

13 Feng Hou and Garnett Picot, "Visible Minority Neighbourhoods in Toronto, Montreal and Vancouver," *Canadian Social Trends*, Spring 2004, 8–13.

14 Mohammad Qadeer, Sandeep K. Agrawal, and Alexander Lovell, "Evolution of Ethnic Enclaves in the Toronto Metropolitan Area, 2001–2006," *Journal of International Migration and Integration* 11 (2010), 324–7.

15 Ibid., 326–7.

16 Ibid., 329.

17 Peach, "The Ghetto and the Ethnic Enclave," 31.

18 Qadeer, Agrawal, and Lovell, "Evolution of Ethnic Enclaves," 315–39, and Mohammad Qadeer and Sandep Kumar, "Ethnic Enclaves and Social Cohesion," *Canadian Journal of Urban Research*, special issue, "Our Diverse Cities: Challenges and Opportunities," vol. 15, no. 2 (2006), 1–17.

19 Daniel Hiebert, Nadine Schuurman, and Heather Smith, *Multiculturalism "on the Ground": The Social Geography of Immigrant and Visible Minority Populations in Montreal, Toronto, and Vancouver, Projected to 2017*, Working paper no. 07–12, Metropolis British Columbia, Centre of Excellence for Research on Immigration and Diversity, April 2007, 8.

20 Map 4.2 is extracted from the map of ethnic communities in New York City. Ford Fessenden and Sam Roberts, "Then as Now – New York's Shifting Ethnic Mosaic," *New York Times*, 22 January 2011, http://www.nytimes.com/interactive/2011/01/23/nyregion/20110123-nyc-ethnic-neighborhoods-map.html?_r=0.

21 New York City Department of City Planning, *The Newest New Yorkers 2000* (New York: Department of City Planning, Population Division, 2004), 86.

22 Nancy Foner, "How Exceptional Is New York? Migration and Multiculturalism in Empire City," *Ethnic and Racial Studies* 30, no. 6 (2007), 12.

23 Nancy Foner, "Introduction: Immigrants in New York City," in *One Out of Three*, ed. Nancy Foner (New York: Columbia University Press, 2013), 17.

24 Changes in the US census have largely shifted socio-economic enumerations to the tri-yearly American Community Surveys. This has reduced the coverage of these topics in the ten yearly census cycles. I have relied on Allen and Turner's analysis and map of ethnic diversity in Southern California for the year 2000. Their gross analysis of 2011 census data for the state suggests that the geographic distribution of ethnic groups continues to follow trends observed in 2000. Thus, Map 4.3 is a reasonable picture of recent ethnic concentrations in Los Angeles

County. James P. Allen and Eugene Turner, *Changing Faces, Changing Places* (Northbridge: University of California, Center for Geographical Studies, 2002), 47, figure 7.1. Also, a personal communication with Eugene Turner regarding state-wide trends in ethnic concentration in 2011 confirmed the trends.

25 James P. Allen and Eugene Turner, "Ethnic Diversity and Segregation in the New Los Angeles," in *EthniCity*, ed. Curtis C. Roseman, Hans D. Laux, and Gunter Thieme (Lanham, MD: Rowman and Littlefield Publishers, 1996), 18.

26 Mike Davis, *Magical Urbanism* (London: Verso, 2000), 64.

27 Allen and Turner, "Ethnic Diversity and Segregation," 46.

28 Min Zhou, Margaret Chin, and Rebecca Kim, "The Transformation of Chinese American Communities," in *New York and Los Angeles*, ed. David Halle and Andrew A. Beveridge (New York: Oxford University Press, 2013), 370.

29 Ibid.

30 Audrey Singer, "An Introduction," in *Twenty-First Century Gateways*, ed. Susan W. Hardwick and Caroline B. Brettell (Washington, DC: Brookings Institution Press, 2008), 3–16.

31 Holger Henke and J.A. George Irish, "Caribbean American Communities in New York City: Perceptions, Conflict and Cooperation," in *Race and Ethnicity in New York City*, ed. Jerome Krase and Ray Hutchison (Amsterdam: Elsevier, 2004), 193–220.

32 Terry Hum, "Immigrant Global Neighborhood in New York City," in *Race and Ethnicity in New York City*, ed. Krase and Hutchison, 25–38.

33 Arun P. Lobo and Joseph J. Salvo, "A Portrait of New York's Immigrant Melange," in *One Out of Three*, ed. Foner, 45.

34 Alejandro Portes and Robert Bach, *Latin Journey: Cuban Immigrants in the United States* (Berkeley: University of California Press, 1985).

35 Brian J.L. Berry, "General Features of Urban Commercial Structure," in *Internal Structure of the City*, ed. Larry S. Bourne (New York: Oxford University Press,1971), 361–7.

36 Shuguang Wang and Jason Zhong, "Delineating Ethnoburbs in Metropolitan Toronto," *CERIS Working Paper no. 100* (Toronto: CERIS, 2013), 20.

37 Mohammad Qadeer and Maghfoor Chaudhry, "The Planning System and the Development of Mosques in the Greater Toronto Area," *Plan Canada* 40, no. 2 (2000), 17–21. Sandeep K. Agrawal, "New Ethnic Places of Worship and Planning Challenges," *Plan Canada*, special edition, "Welcoming Communities" (2009), 64–7.

38 USC Dornsife, crcc.usc.edu/resources/demographics/losangeles.html.
39 Ibid.
40 Richard Sennett, *The Conscience of the Eye* (New York: W.W. Norton and Co., 1990), xiii.
41 Douglas S. Massey and Nancy A. Denton, *American Apartheid: Segregation and the Making of the Underclass* (Cambridge, MA: Harvard University Press,1993), 94.
42 Hiebert, Schuurman, and Smith, *Multiculturalism "on the Ground.*
43 R. Alan Walks and Larry Bourne, "Ghettos in Canada's Cities? Racial Segregation, Ethnic Enclaves and Poverty Concentration in Canadian Urban Areas," *Canadian Geographer* 50, no. 3 (2006), 294–95.
44 Daniel Hiebert, *Exploring Minority Enclave Areas in Montreal, Toronto, and Vancouver*, Citizenship and Immigration Canada, Research and Evaluation Branch (March 2009), 2.
45 Marcuse, "Enclaves Yes, Ghetto No"; Peach, "The Ghetto and the Ethnic Enclave"; Thomas Muller, *Immigrants and the American City* (New York: New York University Press, 1993), 142–60.
46 John R. Logan, Wenquen Zhang, and Richard D. Alba, "Immigrant Enclaves and Ethnic Communities in New York and Los Angeles," *American Sociological Review* 67, no. 2 (2002), 299–322.
47 M. Poulson, J. Forrest, and R. Johnston, "From Modern to Post-Modern? Contemporary Ethnic Residential Segregation in Four U.S. Metropolitan Areas," *Cities* 19, no. 3 (2002), 161–72.
48 Curtis C. Roseman, Hans D. Laux, and Gunter Thieme, "Preface," in *EthniCity*, ed. Roseman et al., xvii.
49 Jude Bloomfield and Franco Bianchini, *Planning for the Intercultural City* (Stroud: Comedia, 2004), 39–40.
50 Mike Savage, Alan Warde, and Kevin Ward, *Urban Sociology, Capitalism and Modernity*, 2nd ed., (Houndmills: Palgrave Macmillan, 2003), 63–7.
51 Allen Scott, *Metropolis: From the Division of Labour to Urban Form* (Berkeley: University of California Press, 1988), 226.
52 Mario Polese and Richard Stren, "Introduction," in *The Social Sustainability of Cities: Diversity and the Management of Change*, ed. Mario Polese and Richard Stren (Toronto: University of Toronto Press, 2000), 3.
53 Ibid., 308–11.

5. Ethnicity and the Urban Economy

1 Ivan Light and Steven Gold, *Ethnic Economies* (San Diego: Academic Press, 2000), 4.

2 Edna Bonancich and John Modell, *The Economic Basis of Ethnic Solidarity: Small Business in the Japanese American Community* (Berkeley: University of California Press,1980), 9.

3 Lucia Lo and Shuguang Wang, "The New Chinese Business Sector in Toronto: A Spatial and Structural Anatomy of Medium-Sized and Large Firms," in *Chinese Ethnic Business, Global and Local Perspectives*, ed. Eric Fong and Chiu Luk (London: Routledge, 2007), 63.

4 Alejandro Portes and Robert L. Bach, *Latin Journey: Cuban and Mexican Immigrants in the United States* (Berkeley: University of California Press, 1985), 203. Light and Gold, *Ethnic Economies*, 23, table 1.3.

5 Light and Gold, *Ethnic Economies*, 24.

6 Leo-Paul Dana, ed., *Handbook of Research on Ethnic Minority Entrepreneurship* (Cheltenham: Edward Elgar, 2007).

7 Light and Gold, *Ethnic Economies*, 19.

8 Giles Barrett, Trevor Jones, and David McEvoy, "Ethnic Minority Business: Theoretical Discourse in Britain and North America," *Urban Studies* 33, nos. 4–5 (1996), 792–3.

9 Robert Kloosterman, Joanne Van Der Leun, and Jan Rath, "Mixed Embeddedness: (In)formal Economic Activities and Immigrant Businesses in the Netherlands," *International Journal of Urban and Regional Research* 23, no. 2 (2001), 257.

10 Richard H. Thompson, *Toronto's Chinatown* (New York: AMS Press, 1989), 220.

11 Min Zhou, *Contemporary Chinese America* (Philadelphia: Temple University Press, 2009), 86–7.

12 Heike Alberts, "Geographic Boundaries of the Cuban Enclave Economy in Miami," in *Landscapes of the Ethnic Economy*, ed. David Kaplan and Wei Li (Lanham, MD: Rowman and Littlefield, 2006), 45.

13 Ibid.

14 Philip Kasinitz, John H. Mollenkopf, Mary C.Waters, and Jennifer Holdaway, *Inheriting the City* (New York: Russell Sage Foundation, 2008), 27.

15 Roger Waldinger, Howard Aldrich, and Robin Ward, "Opportunities, Group Characteristics and Strategies," in *Ethnic Entrepreneurs: Immigrant Businesses in Industrial Societies*, ed. R. Waldinger, H. Aldrich, and R. Ward (London: Sage, 1990), 13–48.

16 Max Weber, *The Protestant Ethic and the Spirit of Capitalism* (1903), trans. Talcott Parsons (Mineola, NY: Dover Publications, 2003).

17 Waldinger, Aldrich, and Ward, "Opportunities, Group Characteristics and Strategies."

18 Thierry Volery, "Ethnic Entrepreneurship: A Theoretical Framework," in *Handbook of Research*, ed. Dana, 33.

19 Waldinger, Aldrich, and Ward, "Opportunities, Group Characteristics and Strategies."

20 Ivan Light and Steven J. Gold, *Ethnic Economies* (San Diego: Academic Press, 2000), 87–91.

21 Ibid., 108–9.

22 Ibid., 127.

23 Eric Fong and Linda Lee, "Chinese Ethnic Economy within the City Context," in *Chinese Ethnic Business*, ed. Fong and Luk, 153–4.

24 Miles Davis, *Magical Urbanism* (London: Verso, 2000), 54.

25 Mohammad Qadeer, *The Bases of Chinese and South Asian Merchants' Entrepreneurship and Ethnic Enclaves, Toronto, Canada*, CERIS Working Paper no. 9 (Toronto: Joint Centre of Excellence for Research on Immigration and Settlement,1999), 28.

26 Michael Ornstein, *Ethno-Racial Groups in Toronto, 1971–2001: A Demographic and Socio-Economic Profile* (Toronto: Institute of Social Research, York University, 2006), table 4.1b.

27 Ibid., table 4.1b.

28 Ibid., table 4.1a.

29 David D. Kallick, "Immigrants and Economic Growth in New York City," in *One Out of Three*, ed. Foner, 76.

30 Ibid., 69.

31 Ibid., 83.

32 Ibid., 84, table 3.

33 New York City Economic Development Corporation, *Bulletin #04-10*, 160 and 165, tables 6.6 and 6.9.

34 Ivan Light and Elizabeth Roach, "Self Employment: Mobility Ladder or Economic Lifeboat?" in *Ethnic Los Angeles*, ed. Roger Waldinger and Mehdi Bozorgmehr (New York: Russell Sage Foundation, 1996),193.

35 US Census Bureau, American Community Survey 2006–8.

36 Light and Roach, "Self Employment," 199.

37 Vilma Ortiz, "The Mexican-origin Population Permanent Working Class or Emerging Middle Class," in *Ethnic Los Angeles*, ed. Waldinger and Bozorgmehr, 260.

38 David Grant, Melvin Oliver, and Angela James, "African Americans: Social and Economic Bifurcation," in *Ethnic Los Angeles*, ed. Waldinger and Bozorgmehr, 406–7.

39 David L. Gladstone and Susan S. Fainstein, "The New York and Los Angeles Economies from Boom to Crisis," in *New York and Los Angeles*, ed.

David Halle and Andrew Beveridge (New York: Oxford University Press, 2013) 95–6.

40 David Hulchanski, "The Three Cities within Toronto: Income Polarization among Toronto's Neighbourhoods, 1970–2000," *Bulletin # 41* (Toronto: University of Toronto, Centre of Urban and Community Studies, 2007).

41 Light and Gold, *Ethnic Economies*, 52.

42 Light and Roach, "Self Employment," 200, table 7.2.

43 Ibid.

44 Ibid., 203, table 7.3.

45 Light and Gold, *Ethnic Economies*, 59–60.

46 Ornstein, *Ethno-Racial Groups in Toronto*, 65.

47 Ibid., table 4.3.

48 Ibid.

49 Jeffery G. Reitz, "Ethnic Concentrations in Labour Market and Their Implications for Ethnic Inequality," in *Ethnic Identity and Equality*, ed. Raymond Breton et al. (Toronto: University of Toronto Press, 1990), 146–95.

50 Kallick, "Immigrants and Economic Growth in New York City," 84, table 3.6.

51 Aljendro Portes and Ruben G. Rumbaut, *A Portrait of Immigrant Americans*, 3rd ed. Berkeley: University of California Press, 2006), 101, figure 5.

52 Ibid., 92–3.

53 Lu Wang, "An Investigation of Chinese Immigrant Consumer Behavior in Toronto, Canada," *Journal of Retailing and Consumer Services* 11 (2004), 317.

54 These numbers are for Chinese of a single ethnic identity. Those of mixed backgrounds are not included.

55 New York City, Department of City Planning, *The Newest New Yorkers* (New York: City of New York, 2013), 20, table 2.5.

56 Lo and Wang, "The New Chinese Business Sector in Toronto," 65.

57 Bonnie Tsui, *American Chinatown* (New York: Free Press, 2009), 118.

58 Thompson, *Toronto's Chinatown*, 14.

59 Ibid., 108.

60 Qadeer, *Bases of Chinese and South Asian Merchants' Entrepreneurship*, 6, 14.

61 Ibid., 36.

62 Yen-Fen Tseng, "Chinese Ethnic Economy: San Gabriel Valley, Los Angeles County," *Journal of Urban Affairs* 16, no. 2 (1994), 172.

63 Yu Zhou, "How Do Places Matter? A Comparative Study of Chinese Economies in Los Angeles and New York City," *Urban Geography* 19, no. 6 (1998), 531–53.

64 Min Zhou, Margaret Chin, and Rebecca Kim, "The Transformation of Chinese American Communities, New York vs. Los Angeles," in *New York and Los Angeles*, ed. D. Halle and A. Beveridge, 373.

65 Lo and Wang, "The New Chinese Business Sector in Toronto," 80.
66 Ibid.
67 Ibid.,154–64.
68 Tsui, *American Chinatown*, 74.
69 Ibid., 178.
70 James P. Allen and Eugene Turner, *Changing Faces, Changing Places* (Northbridge: California State University Press, 2002), 39.
71 Wei Li and Gary Dymski, "Globally Connected and Locally Embedded Financial Institutions," in *Chinese Ethnic Business*, ed. Fong and Luk, 57.
72 Maria W.L. Chee, Gary A. Dymski, and Wei Li, "Asia in Los Angeles," in *Chinese Enterprise, Transnationalism and Identity*, ed. Edmund T. Gomez and Hsin-Huang M. Hsiao (London: Routledge Curzon, 2004), 226.
73 Richard Florida, *The Flight of the Creative Class* (New York: Collins, 2005), 35–6.
74 Kasinitz, Mollenkopf, Waters, and Holdaway, *Inheriting the City*, 348.
75 Light and Gold, *Ethnic Economies*, 70.
76 Ibid., 57.
77 Peter Li, "Ethnic Enterprise in Transition: Chinese Businesses in Richmond B.C. 1980–90," *Canadian Ethnic Studies* 24 (1992), 120–38.
78 Eric Fong and Emi Ooka, "The Social Consequences of Participating in the Ethnic Economy," *International Migration Review* 36, no. 1 (March 2002),125
79 Massachusetts General Hospital in Boston was reported to be electronically outsourcing its readings of X-rays to radiologists in India.
80 Saskia Sassen, *The Global City* (Princeton: Princeton University Press, 1991), 317.
81 Florida, *The Flight of the Creative Class*.
82 Light and Gold, *Ethnic Economies*, 50.
83 Frank A. Butter, Erno Masurel, and Robert H.J. Mosch, "The Economics of Co-ethnic Employment Incentives, Welfare Effects and Policy Options," in *Handbook of Research*, ed. Leo-Paul Dana, 55–6.

6. The Patterns of Community Life

1 Devah Pager and Hana Shepherd offer a comprehensive review of literature on the historical attitudes towards ethno-racial groups. See Devah Pager and Hana Shepherd, "The Sociology of Discrimination: Racial Discrimination in Employment, Housing, Credit, and Consumer Markets," *American Review of Sociology* 34 (January 2008), 81–209. Jeffrey G. Reitz, *The Survival of Ethnic Groups* (Toronto: McGraw-Hill Ryerson, 1980), 4–14.

2 These four aspects are a variation on three areas of multi-ethnic cities identified by Eric Fong and Kumiko Shibuya in "Multiethnic Cities in North America," *Annual Review of Sociology* 31 (2005), 285–304.
3 For references to the works of these theorists see the bibliography.
4 Mike Savage, Alan Warde, and Kevin Ward, *Urban Sociology, Capitalism and Modernity*, 2nd ed., (Houndsmill: Palgrave-Macmillan, 2003), 121.
5 Richard Florida, *The Rise of the Creative Class* (New York: Basic Books, 2002).
6 Manuel Castells, *City, Class and Power* (London: Macmillan Press, 1978), 15–16.
7 Hans Bahrdt, "Public Activity and Private Activity as Basic Forms of City Association," in *Perspectives on the American Community*, ed. Roland Warren (Chicago: Rand McNally & Co., 1966), 78–85.
8 Savage, Warde, and Ward, *Urban Sociology*, 70–105.
9 "Place shapes and constrains our opportunities not only to acquire income but also to become fully functioning members of the economy, society and polity." Peter Drier, John Mollenkopf, and Todd Swanstrom, *Place Matters* (Lawerence: University of Kansas Press, 2004), 28.
10 Philip Kasinitz, John H. Mollenkopf, Mary C. Waters, and Jennifer Holdaway, *Inheriting the City* (New York: Russell Sage Foundation, 2008), 82, figure 2.6.
11 Among immigrants some part of the incidence of one-parent families may be attributed to the temporary separation of parents due to the delayed immigration of a parent, but the differences among immigrant groups are so large that ethnic-cultural factors are unmistakably the main contributors.
12 Kasinitz et al., *Inheriting the City*, 116–18.
13 Bonnie Tsui, *American Chinatown: A People's History of Five Neighborhoods* (New York: Free Press, 2009).
14 Matthew Jacobson, *Whiteness of a Different Color: European Immigrants and the Alchemy of Race* (Cambridge: Harvard University Press, 1998).
15 Christopher B. Doob, *Race, Ethnicity and the American Urban Mainstream* (Boston: Pearson, 2005), 50–7.
16 Douglas Massey and Nancy Denton, "Spatial Assimilation as a Socioeconomic Outcome," *American Sociological Review* 50, no. 1 (1985), 94–106.
17 Ceri Peach, "The Ghetto and Ethnic Enclave," in *Desegregating the City*, ed. David Varady (Albany: State University of New York Press, 2005), 31.
18 John Myles and Feng Hou, "Changing Colours: Spatial Assimilation and New Racial Minority Immigrants," *Canadian Journal of Sociology* 29, no. 1

(2004), 29–55. Eric Fong and Rima Wilkes, "The Spatial Assimilation Model Reexamined: An Assessment of Canadian Data," *International Migration Review* 33, no. 3 (1999), 594–620.

19 Min Zhou, *Contemporary Chinese America* (Philadelphia: Temple University Press, 2009), 6–7.

20 Frederick Boal, "Urban Ethnic Segregation and the Scenario Spectrum," in *Desegregating the City*, ed. Varady, 65.

21 R. Alan Walks and Larry Bourne, "Ghettos in Canada's Cities? Racial Segregation, Ethnic Enclaves and Poverty Concentration in Canadian Urban Areas," *The Canadian Geographer* 50, no. 3 (2006), 294–5.

22 Robert Murdie, "Diversity and Concentration in Canadian Immigration: Trends in Toronto, Montreal and Vancouver 1971–2006," *Research Bulletin #42* (Toronto: University of Toronto, Centre of Urban and Community Studies, 2008). Mohammad Qadeer, Sandeep Agrawal, Alexander Lovell, "Evolution of Ethnic Enclaves in the Toronto Metropolitan Area, 2001–2006," *Journal of International Migration and Integration* 11 (2010), 315–39.

23 David Hulchanski, "The Three Cities within Toronto: Income Polarization among Toronto's Neighbourhoods, 1970–2000," *Bulletin # 41* (Toronto: University of Toronto, Centre of Urban and Community Studies, 2007), 12.

24 Home ownership rates for immigrants catch up to the national level after 16 and more years of living in Canada, with 68.3% ownership for immigrants of 16–25 years' standing compared with a rate of 68.4% for all households. See Canada, CMHC, "2006 Census Housing Series: Issue 7 – The Housing Conditions of Immigrant Households," *Research Highlight, Socio-economic series* 10-016, October 2010, 3, table 3. Some immigrant groups fare much better than others in attaining home ownership. European (non-visible-minority) immigrants had a home ownership rate of 69.7%, while Chinese had the highest rate, 72.4%, compared to the overall rate of 68.4% in 2006. Canada, CMHC, "2006 Census Housing Series: Issue 14 –The Housing Conditions of Visible Minority Households," *Research Highlight, Socio-economic series* 11-009, October 2011, 5, table 3.

25 Daniel Hiebert, *Exploring Minority Enclave Areas in Montreal, Toronto, and Vancouver* (Ottawa: Research and Evaluation Branch, Citizenship and Immigration Canada, 2009), 27.

26 John Goering, ed., "Introduction," in *Fragile Rights within Cities* (Lanham, MD: Rowman & Littlefield, 2007), 12.

27 There is a long-running debate in the literature about the causes of Blacks' social and economic lag. It is between those holding slavery and other societal structures as the cause and those emphasizing family institutions,

habits, and cultural values and norms as the sources of Blacks' poor standing. This debate is not central to our concern.

28 Beveridge et al., "Residential Diversity," 313.
29 Ibid., 318.
30 Robert Murdie and Carlos Teixeira, "Towards a Comfortable and Appropriate Housing: Immigrant Experiences in Toronto," in *The World in a City*, ed. Paul Anisef and Michael Lanphier (Toronto: University of Toronto Press, 2003), 155.
31 Min Zhou, Margaret Chin, and Rebbeca Kim, "The Transformation of Chinese American Communities," in *New York and Los Angeles*, ed. Halle and Beveridge, 365–6, 376–9.
32 John Logan and Weiwei Zhang, "Separate but Equal: Asian Nationalities in the U.S.," in US 2010 Project, http://www.s4.brown.edu/us2010/Data/Report/report06112013.pdf.
33 Ibid.
34 Qadeer, Agrawal, and Lovell, "Evolution of Ethnic Enclaves," 328, table 4.
35 Logan and Zhang, "Separate but Equal."
36 William Frey, "New Projections Point to a Majority Minority Nation in 2044," in *The Avenue/Rethinking Metropolitan America*, 2014, www.brookings.edu.
37 Barry Wellman and Barry Leighton, "Networks, Neighborhoods, and Communities," *Urban Affairs Quarterly* 14, no. 3 (1979), 363–99.
38 Joseph Berger, *The World in a City* (New York: Ballantine Books, 2007), 138.
39 Ibid., 135.
40 Interview with the informant.
41 Claude Fisher, "The Subcultural Theory of Urbanism: A Twentieth-Year Assessment," *American Journal of Sociology* 195, no. 3 (1995), 555.
42 John Lorinc, *The New City* (Toronto: Penguin Canada, 2008), 2–3.
43 Regarding New York's and Los Angeles's images in movies and how they have evolved, see David Halle, Eric Vanstrom, et al., "New York, Los Angeles and Chicago as Depicted in Hit Movies," in *New York and Los Angeles*, ed. Halle and Beveridge.
44 Benson Tong, *The Chinese Americans* (Boulder: University Press of Colorado, 2003), 157.
45 Ibid., 160.
46 The struggle of Muslim women for equality in religious institutions is highlighted by the example of a young mother dealing with a mosque's leadership in Paul Barrett, *American Islam* (New York: Farrar, Straus and Giroux, 2006), 134–78.
47 Tong, *The Chinese Americans*, 236–41.

48 Ibid., 245–8.
49 Some studies of the 1980s show that 31.5% of Chinese marriages involved non-Chinese partners. Quoted ibid., 253.

7. Experiences of Living in Multicultural Cities

1 George Simmel, "The Metropolis and Mental Life," in *Perspectives on the American Community*, ed. Ronald Warren (Chicago: Rand McNally, 1965), 14.
2 Mark Hutter, *Experiencing Cities* (Boston: Pearson, 2007), 10.
3 Richard Sennet, *The Conscience of the Eye* (London: W.W. Norton, 1990), 129.
4 Amanda Wise and Selvaraj Velayutham, eds, "Introduction," in *Everyday Multiculturalism* (Houndmills: Palgrave MacMillan, 2009), 2.
5 Ash Amin, " Ethnicity and the Multicultural City: Living with Diversity," *Environment and Planning A* 34 (2002), 959.
6 Jennifer Lee, *Civility in the City* (Cambridge, MA: Harvard University Press, 2002), 186.
7 Ibid., 3.
8 Chong-Suk Han, "We Both Eat Rice, but That Is about It: Korean and Latino Relations in Multi-Ethnic Los Angeles," in *Everyday Multiculturalism*, ed. Wise and Velayutham (2007), 240–1.
9 Ibid., 239.
10 David Leys, *Millionaire Migrants: Trans-Pacific Lifelines* (London: Wiley Blackwell, 2010).
11 Lucie Cheng and Philip Yang, "Asians: The 'Model Minority' Deconstructed," in *Ethnic Los Angeles*, ed. R. Waldinger and M. Bozorgmehr (New York: Russell Sage, 1996), 305–44. Also Min Zhou, *Contemporary Chinese America* (Philadelphia: Temple University Press, 2009), 221–35.
12 *Maclean's*, "Too Asian," 22 November 2010, 15.
13 The second-generation native-born ethnics are a sizeable group. For example, in 2001, 25% of Chinese, 45% of Blacks, and 65% of Japanese in Canada were Canadian-born. Tina Chui, Kelly Tran, and John Flanders, "Chinese Canadians Enriching the Cultural Mosaic," *Statistics Canada, Catalogue Number 11-006* (Spring 2005), 27.
14 Wise and Velayutham, "Introduction," 4–5.
15 Samuel Beresky, "A Movable Feast," *Planning*, February 2011, 32.
16 City of Toronto, *Street Food Pilot Project Update and Recommendations, Staff Report* (2011), http://www.toronto.ca/legdocs/mmis/2011/ex/bgrd/backgroundfile-37387.pdf, 1–14.
17 Stanley Fish, quoted by Amy Lavender Harris, *Imagining Toronto* (Toronto: Mansfield Press, 2010), 196.

18 *Wikipedia*, "German-American," https://en.wikipedia.org/wiki/German_ American.

19 Min Zhou, *Contemporary Chinese America*, 129.

20 Ibid., 135–7.

21 http://www.rogers.com/web/support/tv/channels/208?setLanguage=en.

22 www.asiantelevision.com/.

23 Richard Florida postulates three Ts of economic growth, namely, talent, technology, and tolerance, tolerance being not a passive act but an active approach of including different people and their perspectives. Florida, *The Flight of the Creative Class* (New York: Collins, 2007), 29–30.

24 Philip Kasinitz, John H. Mollenkopf, Mary C. Waters, and Jennifer Holdaway, *Inheriting the City* (New York: Russell Sage Foundation, 2008), 355.

25 *New York Times*, "India Hitching Hopes on Subway," 4 May 2010, http://www.nytimes.com/glogin?URI=http%3A%2F%2Fwww.nytimes.com%2F2 010%2F05%2F14%2Fworld%2Fasia%2F14delhi.html%3F_r%3.

26 Jane Jacobs, *The Death and Life of Great American Cities* (New York: Random House, 1961; Vintage Books, 1991), 32.

27 Lyn Lofland's views are paraphrased by Hutter, *Experiencing Cities*, 204–5.

28 Ibid., 205.

29 Joe Frienson, "The GTA's Matchless Sports," *Globe and Mail*, 14 August 2011, M1.

30 *New York Times*, "Playing a Sport with Bats and Balls, but No Pitcher," 3 April 2008, http://www.nytimes.com/glogin?URI=http://www.nytimes .com/2008/04/03/nyregion/03cricket.html&OQ=_rQ3D0&OP=690e9591 Q2FfkQ3CcfmucfSSSflcb-fQ5EkHtQ5DkkcgfgQ22Q22WfQ221fQ22VfmuQ 5DrD.kmfQ22VHQ5D.HLrcQ25lcb.

31 Frienson, "The GTA's Matchless Sports."

32 This notice was displayed in front of Monarch Park High School at Christmas time, 2009.

33 Hutter, *Experiencing Cities*, 221.

34 Nicholos Peart (2012), "Why Is N.Y.P.D. After Me," *The New York Times*, December 18, 2011. www.nytimes/2011/12/18/opinion/sunday/young -black-and-frisked-by-the-ypd.html/ref=c (accessed January 19, 2012).

35 Larry Aubry, "Black–Korean American Relations: An Insider's View," in *Los Angeles: Struggles toward Multiethnic Community*, ed. Edward Chang and Russell Leong (Seattle: University of Washington Press, 1993), 149.

36 Ibid., 150.

37 Ella Stewart carried out a study of the perceptions of Korean business owners and African American customers a year after the Los Angeles riots. He found reluctance in both groups about understanding the others' point

of view. Yet they carried on routine dealings with a mixture of courtesy and indifference. Ella Stewart (1993), "Communications Between African Americans and Korean Americans Before and After Los Angeles Riots," in *Los Angeles- Struggles Toward Multiethnic Community*, Edward Chang and Russell Leong, eds., op.cit., 53–54.

38　Philip Napoli, "Little Italy: Resisting the Asian Invasion 1965–96," in *Race and Ethnicity in New York City*, ed. Jerome Krase and Ray Hutchison (Amsterdam: Elsevier, 2004), 248.

39　Ibid., 255.

40　*New York Times*, "In Neighborhood That's Diverse, a Push for Signs to Be Less," 2 August 2011, www.nytimes.com?2011/08/02/nyregion/queens -councilman-wants-english-signs. html.

41　See, for example, Vinay Lal, "Sikh Kirpans in California Schools: The Social Construction of Symbols, Legal Pluralism, and the Politics of Diversity," *Amerasia Journal* 22, no. 1 (1996), 57–89. *Globe and Mail*, "Religion Topples Multiculturalism," 5 October 2010, A10. Gerard Bouchard and Charles Taylor, *Building the Future*, Report, Commission de Consultation sur les Pratiques d'Accommodement Reliées aux Différences Culturelles (Quebec, 2008), 47–55.

42　The term civic culture is derived from the notion of public culture described by Sharon Zukin as involving "shaping public space for social interaction and constructing a visual representation of the city." Sharon Zukin, *The Culture of Cities* (Oxford: Blackwell, 1995), 24.

43　Hutter, *Experiencing Cities*, 390.

44　This sentence draws on Sharon Zukin's idea about public places as sites of public culture. See Zukin, *The Culture of Cities*, 259.

8. Political Incorporation and Diversity

1　*New York Times*, *"Asian New Yorkers Seek Power to Match Numbers,"* 24 June 2011, A18.

2　Ibid.

3　Richard Tindal and Susan Tindal, *Local Government in Canada* (Scarborough, ON: Nelson Thompson Learning, 2000), 3.

4　Peter Drier, John Mollenkopf, and Todd Swanstrom, *Place Matters* (Lawrence: University of Kansas Press, 2004).

5　Kristin Good's indicators of responsiveness include the creation of inclusive images, language accommodation, registration of minorities' preferences, access and equity in service delivery, employment equity, improving intercultural relations, grants to community organizations, etc. Kristin Good, "Patterns in Canada's Immigrant-Receiving Cities and Suburbs," *Policy Studies* 26, nos. 3–4 (2005): 267–8.

6 Ibid.

7 John Mollenkopf and Raphael Sonenshein, "The New Urban Politics of Integration: A View from the Gateway Cities," in *Bringing Outsiders In*, ed. Jennifer Hochschild and John Mollenkopf (Ithaca: Cornell University Press, 2009), 74–93.

8 Good, "Patterns in Canada's Immigrant-Receiving Cities," 267–8.

9 A narrow view of political incorporation refers only to how an immigrant or minority group finds a place in a political structure. For a comprehensive account see Jennifer Hochschild and John Mollenkopf. "Modeling Immigrant Political Incorporation," in *Bringing Outsiders In*, 16.

10 Jennifer Hochschild and John Mollenkopf, "Understanding Immigrant Political Incorporation through Comparison," in *Bringing Outsiders In*, 303–4.

11 This paragraph draws on John Mollenkopf and Raphael Sonenshein, "The New Urban Politics of Integration," in *Bringing Outsiders In*, 75–7.

12 Ibid., 91.

13 John Mollenkopf and Raphael Sonenshein, "New York and Los Angeles: Government and Political Influence," in *New York and Los Angeles*, ed. David Halle and Andrew Beveridge (New York: Oxford University Press, 2013), 145, table 5.5.

14 Ibid.

15 Mollenkopf and Sonenshein, "The New Urban Politics of Integration," in *Bringing Outsiders In*, 75.

16 Patricia Pessar and Pamela Graham, "Dominicans: Transnational Identities and Local Politics," in *New Immigrants in New York*, ed. Nancy Foster (New York: Columbia University Press, 2001), 264.

17 Mollenkopf and Sonenshein, "The New Urban Politics of Integration," in *Bringing Outsiders In*, 76.

18 Ibid.

19 While isolated examples of Black councillors elected in Los Angeles can be traced back to 1915, a consistent pattern of Blacks elected councillors started after 1965. John H. Laslett, "Historical Perspectives: Immigration and the Rise of a Distinctive Urban Region, 1900–1970," in *Ethnic Los Angeles*, ed. Roger Waldinger and Mehdi Bozorgmehr (New York: Russell Sage Foundation, 1996), 68.

20 Mollenkopf and Sonenshein, "The New Urban Politics of Integration," in *Bringing Outsiders In*, 75.

21 Mollenkopf and Sonenshein, "New York and Los Angeles," in *New York and Los Angeles*.

22 The Canadian literature on the politics of inclusion is largely, though not exclusively, conceived in terms of how local and national institutions are responding to the minorities' needs, namely, what institutions are doing for ethno-racial and immigrant groups and not so much how they are gaining their fair share in political power. See Good, "Patterns in Canada's Immigrant-Receiving Cities"; Frances Frisken and Marcia Wallace, "Governing the Multicultural Region," *Canadian Public Administration* 46, no. 2 (2003), 153–77; and Katherine Graham and Susan Phillips, "Another Fine Balance: Managing Diversity in Canadian Cities," in *Belonging?* ed. Keith Banting, Thomas Courchene, and F. Leslie Seidle (Montreal: Institute for Research on Public Policy, 2007),155–95. There are others who address the question of political representation of minorities. See Caroline Andrew, John Biles, Myer Siemiatycki, and Erin Tolly, eds, *Electing a Diverse Canada: The Representation of Minorities and Women* (Vancouver: UBC Press, 2008) and Myer Siemiatycki and Anver Saloojee, "Ethno-racial Political Representation in Toronto: Patterns and Problems," *Journal of International Migration and Integration* 3, no. 2 (2002), 241–73.

23 Siemiatycki and Saloojee, "Ethno-racial Political Representation in Toronto," 242 and Caroline Andrew, John Biles, Myer Siemiatycki, and Erin Tolly, "Introduction," in *Electing a Diverse Canada*, ed. Andrew, Biles, Siemiatycki, and Tolly, 5.

24 Nominally, the English monarch appoints Canadian governors generals, but she/he takes the recommendation of the prime minister.

25 The GTA is slightly bigger than the Census Metropolitan Area (CMA) of Toronto. It is essentially a conceptual aggregation of municipalities that are closely tied as an economic and labour market, without any corporate structure.

26 Siemiatycki and Saloojee, "Ethno-racial Political Representation in Toronto," 249, table 1.

27 Myer Siamiatycki, *The Diversity Gap:The Electoral Under-representation of Visible Minorities*, (Toronto: Ryerson Centre of Immigration and Settlement, 2011), 12.

28 Ibid., 12.

29 In the 2014 municipal elections in the Toronto area, including suburbs, numerous ethnic candidates contested for local councillors' and mayors' positions. For the two largest ethnic groups, Chinese and South Asians, candidates of such backgrounds were in the majority in Brampton and Mississauga, and in their ethnic enclaves of Toronto city and Markham. Yet they were largely unsuccessful, trailing far behind the winners. For example, in Brampton, a predominantly South Asian suburb, only one out

of ten elected councillors was South Asian, despite a majority of candidates being of similar background. The results bear out the fact that ethnic voters looked for the personal qualities, political sophistication, and public-service record of their preferred representatives and showed little ethnic loyalty.

30 Myer Siemiatycki, "Reputation and Representation: Reaching for Political Inclusion," in *Electing a Diverse Canada*, ed. Andrew, Biles, et al., 35.

31 Siemiatyki, *The Diversity Gap*, 8.

32 Irene Bloemraad, "Diversity and Elected Officials in the City of Vancouver," in *Electing a Diverse Canada*, ed. Andrew, Biles, et al., 56. It may be noted that in the 2005 Vancouver city council elections, two additional Chinese councillors were elected (ibid., 68 n. 13).

33 Geetika Bagga, "From the Komagata Maru to Six Sikh MPs in Parliament," *Our Diverse Cities*, no. 4 (2007), 161–5.

34 That representatives pursue objectives on the basis of group identities and interests is a collectivist view of the political arena. This idea is contrasted with the liberal individualist model of representation. Christopher Anderson and Jerome Black, "The Political Integration of Newcomers, Minorities and Canadian-born: Perspectives on Naturalization, Participation, and Representation," in *Immigration and Integration in Canada in the Twenty-First Century*, ed. John Biles, Meyer Burstein, and James Frideres, (Montreal: School of Policy Studies, Queen's University, McGill-Queen's University Press, 2008), 60–1.

35 Myer Siemiatycki and Anver Saloojee, "Ethno-racial Political Representation in Toronto," 241–73.

36 Ibid., 267 and 268.

37 Ibid., 268.

38 Bloemraad, "Diversity and Elected Officials in the City of Vancouver," 66.

39 Peter Burns, *Electoral Politics Is Not Enough* (Albany: State University of New York Press, 2006), 4–5.

40 Ibid.,100–3.

41 Ibid., 4–5.

42 Before the Voting Rights Act (1965), there were on average only 300 elected Black representatives at all levels of the US government. In 2011 there were 9000 such elected representatives, including 43 members of Congress and, to crown it all, a Black president. John Lewis, "A Poll Tax by Another Name," *New York Times*, 27 August 2011, http://www.nytimes.com/2011/08/27/opinion/a-poll-tax-by-another-name.html?_r=0.

43 Vancouver housing prices rose 12% in 2010 thanks to mainland Chinese investment. *Globe and Mail*, "Vancouver Cheers Chinese Driven Real Estate Boom," 12 March 2011, A4.

44 C Wright Mills, *The Power Elite* (London: Oxford University Press, 1956).
45 This formulation is a restatement of Clarence Stone, "Urban Regimes and Capacity to Govern: A Political Economy Approach," *Journal of Urban Affairs* 15, no. 1 (1993), 1–28.
46 Good, "Patterns in Canada's Immigrant-Receiving Cities," 276.
47 Ibid., 269–71.
48 Susan Fainstein, *The Just City* (Ithaca: Cornell University Press, 2010).
49 Mike Davis, *Magical Urbanism* (London: Verso, 2000).
50 Joseph Berger, *The World in a City* (New York: Ballantine Books, 2007), viii.

9. The Pluralism of Urban Services

1 William Wilson, *The Truly Disadvantaged: The Inner City, the Underclass, and Public Policy* (Chicago: University of Chicago Press,1987). Jeffery Reitz, *The Survival of Ethnic Groups* (Toronto: McGraw-Hill Ryerson, 1980).
2 In Canada non-White households, who mostly lived in metropolitan areas, were nearly twice as likely to be living under conditions of unmet core (affordability, suitability, and adequacy) housing needs in 2006. Canada Mortgage and Housing Corporation, "Research Highlight," Socio-economic series 08-016 (December 2008). For a historical account the social disparities of minorities in US cities see David Ward, "The Emergence of Central Immigrant Ghettos in American Cities: 1840–1920," *Annals of the Association of American Geographers* 68, no. 2 (1968).
3 Cheryl Huber, "New York Wants to Engage Immigrants in Its Parks," New Yorkers for Parks, Canada.metropolis.net/pdfs/fow_30aug10_huber_e .pdf.
4 Sarah V. Wayland, *Integration of Peel Immigrant Discussion Paper* (Brampton: Region of Peel, Human Services, 2010), 18.
5 Interview with the director of community grant programs, City of Toronto, December 2010.
6 Cecile Poirier, Annick Germain, and Amelie Billette, "Diversity in Sports and Recreation: A Challenge or an Asset for Municipalities of Greater Montreal?" *Canadian Journal of Urban Research*, special issue, vol. 15, no. (2006, supplement), 40–3.
7 James Green, *Cultural Awareness in Human Services* (Boston: Allyn and Bacon, 1999), 87.
8 Mitchell Rice, *Diversity and Public Administration* (Armonk, NY, M.E. Sharpe, 2010), 176.
9 Paraphrased from Green, *Cultural Awareness in Human Services*, 87.
10 Ibid., 110.

11 New York City's information line greets callers in six languages.

12 City of New York, Independent Budget Office, *Letter to the Office of the Minority Leader*, 18 September 2009.

13 International banks on the main street of Toronto have big posters in windows announcing ease of service in various language. Banks in the Chinatowns of New York, Los Angeles, and Toronto have their deposit and withdrawal forms printed in Chinese.

14 Particularly in human services, being served by someone of the same background increases clients' trust. In the counselling professions, however, clients apprehensive about "losing face" in their communities may sometimes prefer counsellors of different backgrounds.

15 It can be argued that ethno-specific agencies isolate their clients from others and thereby contribute to segregation.

16 Kristin Good, "Patterns of Politics in Canada's Immigrant-receiving Cities and Suburbs," *Policy Studies* 26, no. 3/4 (2005), 281.

17 These examples are drawn from actual cases in Montreal, Toronto, New York, and other cities.

18 New York Times, "Little-Known Guide Helps Police Navigate a Diverse City," 10 June 2013, http://www.nytimes.com/2013/06/11/nyregion/a -not-for-tourists-guide-to-navigate-a-multicultural-city-html?

19 Gerard Bouchard and Charles Taylor, *Building the Future*, Report, Commission de Consultation sur les Pratiques d'Accommodement Reliées aux Différences Culturelles (Quebec, 2008).

20 Karen Matthews, "New York City Public Schools to Close Doors on Two Muslim Holidays," *Globe and Mail*, 4 March 2015, http://www .theglobeandmail.com/news/world/new-york-city-public-schools-to -close-doors-on-two-muslim-holidays/article23292118/.

21 Policy of the Toronto District School Board, 2002, no. B.03, pp.77–8.

22 Supreme Court of Canada, *Multani v. Commission Scolaire Marguerite-Bourgeoys*, [2006] S.C.C. 6, 1 S.C.R. 256, and Vinay Lal (1996) "Sikh Kirpans in California Schools: The Social Construction of Symbols, Legal Pluralism, and the Politics of Diversity," *Amerasia Journal* 23, no. 1 (1996), 57–89.

23 There is a vast literature about the historic discrimination against minorities in jobs, housing, and services and the evidence of their social exclusion continues to be reported. See Michael Ornstein, *Ethno-racial Groups in Toronto 1971–2001: A Demographic and Social-economic Profile* (Toronto: Institute for Social Research, York University, 2006) and Camille Zubrinsky Charles, *Won't You Be My Neighbor?* (New York: Russell Sage Foundation, 2006).

24 Bill Mears, "Michigan's Ban on Affirmative Action Upheld by Supreme Court," *CNN News*, 22 April 2014, www.cnn.com/2014/04/22/justice/scotus-michigan-affirmative-action/.
25 Adam Liptak, "Supreme Court Finds Bias against White Firefighters," *New York Times*, 29 June 2009, www.nytimes.com/2009/06/30/us/30scotus.html.

10. Urban Planning for Cultural Diversity

1 These goals of planning are not always compatible in all situations. There is much debate about not only their mutual trade-offs, but also their relative significance. These are treated as generic values whose application is negotiated on the basis of situations. See Gerald Hodge and David Gordon, *Planning Canadian Communities* (Scarborough, ON: ITP/Nelson, 2007), 112–16.
2 Public interest is a much discussed term. One conception holds it to be the common interests or shared notions of welfare held by persons in a political community. An alternative view is that it is the aggregate of individual interests affected by a policy or action. The latter notion is utilitarian in its origins and ends up with the logic of aggregating benefits over costs. These collectivist or communitarian versus individualistic and utilitarian notions have been the source of a long-running debate in theories of public interest. Still, the idea that public interest exists over and above individual goals is enduring. See Iain McLean and Alistair McMillan, eds, *Concise Oxford Dictionary of Politics* (Oxford: Oxford University Press, 2003), 448.
3 Michael Burayidi, "Urban Planning as a Multicultural Canon," in *Urban Planning in a Multicultural Society*, ed. Michael Burayidi (Westport, CT: Praeger, 2000), 14. Leonie Sandercock, *Cosmopolis II* (London: Continuum, 2003), 40–3. Leonie Sandercock and Ann Forsyth, "A Gender Agenda: New Directions for Planning Theory," *Journal of the American Planning Association* 58, no. 1 (1992), 49–59.
4 Patsy Healy, *Collaborative Planning* (Vancouver: UBC Press, 1997), 297.
5 Martin Anderson, *The Federal Bulldozer: A Critical Analysis of Urban Renewal 1949–62,* (Cambridge, MA: MIT Press, 1964).
6 *Rosenberg v. Outremont (City)*, Quebec Superior Court, 6 September 2001. https://www.cardus.ca/lexview/article/2309/.
7 McLean and McMillan, *Concise Oxford Dictionary of Politics*, 175.
8 Planning theory is a term applied to procedural models of planning decision making, namely, the processes of planning decision making,

values that inform the decision makers and the involved actors. It is not a systemized body of knowledge but a collection of propositions, concepts, critiques, ethics, and issues. It does not normally refer to theories of the substantive aspects of urban planning such as land use, housing, or transportation. For an overview of the focus and domain of planning theory, see Ernest Alexander, *Introduction to Current Theories, Concepts and Issues* (New York: Gordon and Branch Science Publishers, 1986).

9 Sandercock, *Cosmopolis II*, 45–6.

10 Reeves, *Planning for Diversity*,11.

11 Katherine Pestieau and Marcia Wallace, "Challenge and Opportunities for Planning in the Ethno-Culturally Diverse City: A Collection of Papers – Introduction," in *Planning Theory and Practice* 4, no. 3 (2003), 253–8. Burayidi, "Urban Planning as a Multicultural Canon," 1–14. Mohammad Qadeer, "Urban Planning and Multiculturalism, Beyond Sensitivity," *Plan Canada* 40, no. 4 (2000), 16–18.

12 Huw Thomas, "Race Equality and Planning: A Changing Agenda," *Planning, Practice and Research* 23, no. 1 (2008), 3.

13 Sandercock, *Cosmopolis II*, 68–82. Michael Buryadi, "Tracking the Planning Profession: From Monistic Planning to Holistic Planning for a Multicultural Society," in *Urban Planning in a Multicultural Society*, ed. Burayidi, 39–51.

14 Sandercock, *Cosmopolis II*, 64.

15 Beth Moore Milroy and Marcia Wallace , *Ethno-racial Diversity and Planning Practices in the Greater Toronto Area* (Toronto: Centre of Excellence for Research in Immigration and Settlement, University of Toronto, 2002), 27.

16 Leela Viswanathan, "Postcolonial Planning and Ethnoracial Diversity in Toronto: Locating Equity in a Contemporary Planning Context," *Canadian Journal of Urban Research* 18, no. 1 (2009), 162–82.

17 Thomas, "Race Equality and Planning: A Changing Agenda," 1.

18 Leonie Sandercock, "Towards a Planning Imagination for the 21st Century," *Journal of the American Planning Association* 70, no. 2 (2004), 140.

19 Ibid., 139.

20 Ibid., 136–7.

21 The model of urban planning in this literature is almost entirely conceived in terms of the planning process. It is evident in the writings of all the theorists cited in this section.

22 Domenic Vitiello, "The Migrant Metropolis and American Planning" *Journal of the American Planning Association* 25, no. 2 (2009), 251.

23 Paul Davidoff, "Advocacy and Pluralism in Planning," *Journal of the American Institute of Planners* 31, no. 4 (1965), 331–8.

24 For a description of current planning approaches see Healey, *Collaborative Planning*, and Judith Innes, "Information in Communicative Planning," *Journal of the American Planning Association* 64, no. 1 (1998), 52–63.

25 Michael P. Brooks, *Planning Theory for Practitioners* (Chicago: Planners Press, 2002), 21.

26 Richard E. Klosterman, "Planning Theory Education: Thirty Years Review," *Journal of Planning Education and Research* 31, no. 3 (2011), 325–6.

27 Gerald Hodge and David Gordon, *Planning Canadian Communities* (Toronto: Nelson, 2007).

28 Community involvement is now a legislated part of the decision-making process in urban planning in both the United States and Canada. Ethno-racial minorities' participation is specifically promoted in urban planning.

29 For example, the competition for political influence between American Blacks and West Indian Blacks played out in the election in New York's congressional district 11 in 1999. Another example is the Chinese community's opposition to the use of a mosque in Markham, Ontario, for funeral prayers and the storing of dead bodies. Such inter-ethnic differences are beginning to surface in multicultural cities.

30 This was a mailed questionnaire survey of the urban planning departments in both countries. It is based on a purposive and not random sample. Yet it is the only countrywide survey of multicultural planning policies and practices in both countries. For the scope, methodology, and findings of this survey see Mohammad Qadeer and Sandeep Agrawal, "The Practice of Multicultural Planning in American and Canadian Cities," *Canadian Journal of Urban Research* 20, no. 1 (supplement, 2011), 132–56.

31 Kristin Good, "Patterns of Politics in Canada's Immigrant-receiving Cities and Suburbs," *Policy Studies* 26, nos. 3–4 (2005), 269.

32 "Muslim Community Centre Park 51," *New York Times* [2011], http://topics.nytimes.com/top/reference/timestopics/organizations/p/park51/index.html.

33 Phil Wilson, "Planned Temecula Valley Mosque Draws Opposition," *Los Angeles Times*, 18 July 2010, articles.latimes.com/2010/jul/18/local/la-me-mosque-20100718.

34 There is no official census of places of worship. These numbers are estimates based on counts of the places of worship registered voluntarily with a website. Many storefront places of worship can be opened without any approval under existing zoning laws. These are seldom registered. Only about 10% of mosques, etc. are built anew and thus appropriately.

35 One recent example is a proposal to convert a vacant Catholic convent into a mosque in Staten Island, New York. The plan had to be abandoned

on account of the fierce opposition mobilized by a city-wide conservative group. Yet only a few blocks from the site, in the Dongan Hills neighbourhood, a mosque was established in a house previously used as a mandir after neighbours were informed and their consent obtained. In Staten Island, there are half a dozen mosques. So episodic is the politics of mosque development. See Joseph Berger, "Protests of a Plan for a Mosque? That Was Last Year," *New York Times*, 19 August 2011, http://www .nytimes.com/2011/08/19/nyregion/mosque-opens-quietly-on-staten -island.html.

36 There are many cases of city leaders striking deals between opponents and sponsors of a proposal to help develop places of worship and satisfy objectors. In the city of Markham (Toronto area), a proposed Hindu mandir was fiercely opposed by the surrounding neighbourhood. In 1993 the mayor negotiated a deal whereby the site of the mandir was rezoned to allow the Hindu community to sell it at a profit and from those proceeds to buy a large piece of land on the periphery for building the mandir. It took many years, yet the mandir eventually developed. Recently mosques, churches, and temples have been established in malls fallen vacant from changing commercial patterns. Both the mall owners and local leadership have combined to reach a resolution. See Marc Hequet, "God and Mall," *Planning*, July 2010, 13–15.

37 Mohammad Qadeer and Maghfoor Chaudhry, "The Planning System and the Development of Mosques in the Greater Toronto Area," *Plan Canada* 40, no. 2 (2000), 17–20.

38 Stacy Harwood, "Struggling to Embrace Difference in Land-Use Decision-Making in Multicultural Communities," *Planning, Practice and Research* 20, no. 4 (2005), 363–4.

39 Sandeep Agrawal, "New Ethnic Places of Worship and Planning Challenges," *Plan Canada*, "Welcoming Communities" special issue, 2009, 64–7.

40 Mike Davis, *City of Quartz* (London: Verso, 1990), 203–8.

41 Ibid., 207–8.

42 American Institute of Certified Planners, "AICP Code of Ethics and Professional Conduct" (revised, 2009), https://www.planning.org/ethics/ ethicscode.htm. Ontario Professional Planners Institute, *Professional Code of Practice* (2006), 1.

43 In 2012, thirteen cities, counties, and other governmental entities in California filed for bankruptcy. These may be extreme cases, but many local governments in both the United States and Canada were in financial difficulties. Under such conditions, urban planning is severely strained to

meet the demands of various communities. For the financial troubles of California cities see *Toronto Star*, "Stockton is Biggest City in U.S. to File for Bankruptcy," 28 June 2012, A32.

44 Amin, *Ethnicity and the Multicultural City*.

45 Ontario Municipal Board, *Tilzen Holdings Ltd versus Town of Markham* (1998), O.M.B.D. no. 319, File no. PL 956623.D950049, Z960134, M970061.

46 Mohammad Qadeer, "Pluralistic Planning for Multicultural Cities: The Canadian Practice," *Journal of the American Planning Association* 63, no. 4 (Autumn 1997), 487.

47 "LA Mayor Signs Law to Limit McMansions in Hillsides," www .catadjuster.org/Forums/tabid/60/aft/11759/Default.aspx.

48 Author's interview with town officials in October 2009.

49 Qadeer, "Pluralistic Planning for Multicultural Cities," 487.

50 Shuguang Wang and Jason Zhong, *Delineating Ethnoburbs in Metropolitan Toronto*, CERIS Working paper no.100 (Toronto: CERIS – Ontario Metropolis Centre, 2013), 20.

51 Quoted in Christopher Hawthorne, "'Latino Urbanism' Influences a Los Angeles in Flux," *Los Angeles Times*, 6 December2014.

52 Jude Bloomfield and Franco Bianchini, *Planning for the Intercultural City* (Stroud: Comedia, 2004), 78–84.

53 Jan Lin, *The Power of Urban Ethnic Places* (New York: Routledge, 2011), 198–204, 210–17.

54 *Globe and Mail*, "The Rise and Fall of the Ethnic Mall," 16 June 2012, M1 and M5.

55 For Toronto's poor neighbourhoods see David Hulchanski, *The Three Cities within Toronto: Income Polarization among Toronto's Neighbourhoods 1970–2000*, Working paper 41 (Toronto: Centre of Community and Urban Studies, University of Toronto, 2007).

56 John Lorinc, "The New Regent Park," *UofTMagazine*, Spring 2013, http:// magazine.utoronto.ca/feature/new-regent-park-toronto-community -housing-john-lorinc/.

57 Sharifa Pitts, *Harlem Is Nowhere* (New York: Little, Brown and Co., 2011), 162–5.

58 John McCarron, "Majoring in Development," *Planning*, October 2011, 13.

59 Leonie Sandercock, "Integrating Immigrants: The Challenge for Cities, City Governments and the City-Building Professions," in *The Intercultural City*, ed. Phil Wood and Charles Landry (London: Earthscan, 2008), 258–72.

60 Yasmeen Abu-Lahan and Baha Abu-Lahan, "Reasonable Accommodation in a Global Village," *Policy Options* 28, no. 8 (2007), 30.

61 Kathleen Weil, Minister of Justice, Province of Quebec, "Bill 94. An Act to Establish Guidelines Governing Accommodation Requests within the

Administration and Certain Institutions" (Quebec Official Publisher, 2010), 1–4. Also, "US legal definition, reasonable accommodation law and legal definition," http://definitions.uslegal.com/r/reasonable -accomodation/.

62 Julius Grey, (2007), "The Paradoxes of Reasonable Accommodation," *Policy Options*. Volume 28, no. 3, (2007): 34–36.

63 Mohammad Qadeer, "What Is This Thing Called Multicultural Planning?" *Plan Canada*, "Welcoming Communities: Planning for Diverse Populations" (Special edition, 2009), 10–13.

64 Gerard Bouchard and Charles Taylor, *Building the Future*, Report, Commission de Consultation sur les Pratiques d'Accommodement Reliées aux Différences Culturelles (Quebec, 2008), 19, 89.

65 For the details and methodology of the survey see Mohammad Qadeer and Sandeep Agrawal, "The Practice of Multicultural Planning in American and Canadian Cities," *Canadian Journal of Urban Research* 20, no. 1 (supplement, 2011), 132–56. The survey sample was purposive and not statistically representative.

66 Ibid., 143.

67 Ibid., 146.

68 Ibid.

69 In these times of the Internet, Googling any city's ethno-racial associations, centres, and community organizations results in hundreds and hundreds of hits.

70 Toronto's official plan, prepared in 2002, was consolidated after various amendments ordered by the Ontario Municipal Board up to 2010. It is available at http://www1.toronto.ca/static_files/CityPlanning/PDF/chapters1_5_dec2010.pdf.

11. Imagining Multicultural Cities

1 Diana Eck, "What Is Pluralism," Pluralism Project at Harvard, http://pluralism.org/pluralism/what_is_pluralism.

2 Pico Iver, *The Global Soul* (New York: Vintage Books, 2000), 123.

3 Myer Siemiatycki, Tim Rees, and Khan Rahi, "Integrating Community Diversity in Toronto: On Whose Terms?" in *The World in a City*, ed. Paul Anisef and Michael Lanphier (Toronto: University of Toronto Press, 2003), 455.

4 Nathan Glazer, *We Are All Multiculturalists Now* (Cambridge: Harvard University Press, 1997).

5 Quoted in Hilton Als, "Playing to Type," *The New Yorker*, 23 May 2011, 86.

6 Nancy Foner, "How Exceptional Is New York? Migration and Multiculturalism in the Empire City," *Ethnic and Racial Studies* 30, no. 6 (2007), 1015.

7 Philip Kasinitz, John H. Mollenkopf, and Mary C. Waters, *Becoming New Yorkers: Ethnographies of the New Second Generation* (New York: Russell Sage Foundation, 2004),16.

8 John Hull Mollenkopf, "School Is Out," in *The City Revisited*, ed. Dennis R. Judd and Dick Simpson (Minneapolis: University of Minnesota Press, 2011), 169–71.

9 Michael Dear and Nicholas Dahmann, "Urban Politics and the Los Angeles School of Urbanism," in *The City Revisited*, ed. Judd and Simpson, 66–8.

10 Terry Nichols Clark, "The New Chicago School," in *The City Revisited*, ed. Judd and Simpson, 235.

11 Ibid., 230.

12 Joel Kotkin, "City of Villages," *City- Journal*, Winter 2014.

13 Herbert J. Gans, quoted by Foner, "How Exceptional Is New York?" 1016.

14 Ibid., 1005–7.

15 Will Kymlicka, *Multicultural Odysseys* (New York: Oxford University Press, 2007), 83.

16 Jan Rath, "Debating Multiculturalism: Europe's Reaction in Context," *Harvard International Review*, 6 January 2011, http://hir.harvard.edu/archives/2773.

17 Ibid.

18 Doug Saunders, *Arrival City* (Toronto: Alfred A. Knopf Canada, 2010).

19 Dowell Myers, *Immigrants and Boomers* (New York: Russell Sage Foundation, 2007), 155. He draws on Beth A. Rubin, *Shifts in the Social Contract: Understanding Change in American Society* (Thousand Oaks, CA: Sage Publications, 1996).

20 Ibid., 252–3.

Appendix

1 Paraphrased from Peter Kivisto, "We Really Are All Multiculturalists Now," *Sociological Quarterly* 5, no. 3.1 (2012), 4.

2 In 1956 Horace Kallen wrote *Cultural Pluralism and the American Idea: An Essay in Social Philosophy* (Philadelphia: University of Pennsylvania Press, 1956). Also cultural pluralism as the characterization of New York City's society was advanced by Nathan Glazer and Daniel Moynihan in *Beyond the Melting Pot* (Cambridge, MA: MIT and Harvard University Presses, 1965).

3   Angie Fleras, *The Politics of Multiculturalism* (New York: Palgrave, 2009), 14.

4   Ibid.,15.

5   Ibid.,17.

6   Will Kymlicka, *Multicultural Odysseys* (New York: Oxford University Press, 2007), 73.

7   Howard Duncan, "Multiculturalism: Still a Viable Concept for Integration," *Canadian Diversity* 4, no. 1 (Winter 2005), 13.

8   Kymlicka, *Multicultural Odysseys*, 167.

9   Will Kymlicka, "Ethnocultural Diversity in a Liberal State: Making Sense of the Canadian Model(s)," in *Belonging? Diversity, Recognition and Shared Citizenship in Canada*, ed. Keith Banting, Thomas Courchene, and F. Leslie Seidle (Montreal: Institute of Public Policy, 2007), 54.

10   Jude Bloomfield and Franco Bianchini, *The Intercultural City* (Stroud: Comedia, 2004), 39.

11   Gerard Bouchard and Charles Taylor, *Building the Future*, Report of Commission de Consultation sur les Pratiques d'Accommodement Reliées aux Différences Culturelles (Quebec, 2008), 19.

12   Banting, Courchene, and Seidle, "Introduction" in *Belonging?* ed. Banting, Courchene, and Seidle, 1.

13   Neil Bissoondath, *Selling Illusions* (Toronto: Penguin, 1994), 113.

14   David Hollinger, *Postethnic America: Beyond Multiculturalism* (New York: Basic Books, 2000), 3.

15   David Brooks, "The Death of Multiculturalism," *New York Times*, 27 April 2006, http://query.nytimes.com/gst/fullpage.html?res=9C01E1DC133FF9 34A15757C0A9609C8B63.

16   Lawrence Martin, "Enough of Multiculturalism – Bring on the Melting Pot," *Globe and Mail*, 31 March 2009, A17.

17   Jan Rath, "Debating Multiculturalism," *Harvard International Review*, 6 January 2011, http://hir.harvard.edu/archives/2773.

18   Arjun Appadurai, *The Future as a Cultural Fact* (London: Verso, 2013), 179, 182.

19   Kian Tajbakhsh, *The Promise of the City* (Berkeley: University of California Press, 2001), 164.

20   Bhikhu Parekh, quoted in Ziauddin Sardar, *Balti Britain* (London: Granta, 2008), 352.

Bibliography

Abu-Laban, Yasmeen, and Baha Abu-Laban. "Reasonable Accommodation in a Global Village." *Policy Options* 26, no. 8 (September 2007): 30–3.

Access Alliance. "Racialised Groups and Health Status." http://accessalliance .ca/wp-content/uploads/2015/03/Racialised_Groups_Health_Status _Literature_Review.pdf.

Adams, Michael. *Unlikely Utopia.* Toronto: Viking, 2007.

Agrawal, Sandeep K. "New Ethnic Places of Worship and Planning Challenges." *Plan Canada, Welcoming Communities.* Special issue, 2009, 64–7.

Agrawal, Sandeep K., Mohammad A. Qadeer, and Arvin Prasad. "Immigrants' Needs and Public Service Provisions in Peel Region." *Plan Canada* 47, no. 2. (Summer 2007), 48–50.

Alba, Richard, and Victor Nee. *Remaking the American Mainstream.* Cambridge, MA: Harvard University Press, 2003.

Alberts, Heike. "Geographic Boundaries of the Cuban Enclave Economy in Miami." In *Landscapes of the Ethnic Economy*, edited by David H. Kaplan and Wei Li, 35–48. Lanham: Rowman and Littlefield, 2006.

Alexander, Ernest. *Introduction to Current Theories, Concepts and Issues.* New York: Gordon and Branch Science Publishers, 1986.

Allen, James P., and Eugene Turner. *Changing Faces, Changing Places.* Northridge: University of California, Center for Geographical Studies, 2002.

Allen, James P., and Eugene Turner. " Ethnic Diversity and Segregation in the New Los Angeles." In *EthniCity*, edited by Curtis C. Roseman, Hans D. Laux, and Gunter Thieme, 11–23. London and Lanham, MD: Rowman and Littlefield Publishers,1996.

Als, Hilton. "Playing to Type." *The New Yorker*, 23 May 2011.

American Institute of Certified Planners. "AICP Code of Ethics and Professional Conduct." Revised, 2009. https://www.planning.org/ethics/ethicscode.htm.

Amin, Ash. "Ethnicity and the Multicultural City, Living with Diversity." *Environment and Planning A* 34 (2002).

Amin, Ash, and Nigel Thrift. *Cities: Reimagining Urban.* Cambridge: Polity, 2002.

Amin, Ash. "The Good City." *Urban Studies* 43, nos. 5/6 (May 2006).

Anderson, Christopher, and Jerome Black. "The Political Integration of Newcomers, Minorities and Canadian-Born: Perspectives on Naturalization, Participation, and Representation." In *Immigration and Integration in Canada in the Twenty-First Century, edited by* John Biles, Meyer Burstein, and James Frideres 45–76. Kingston, ON, and Montreal: Queen's University School of Policy Studies, McGill-Queen's University Press, 2008.

Anderson, Martin. *The Federal Bulldozer: A Critical Analysis of Urban Renewal, 1949–62.* Cambridge, MA: MIT Press,1964.

Andrew, Caroline, John Biles, Myer Siemiatycki, and Erin Tolly, eds. *Electing a Diverse Canada: The Representation of Minorities and Women.* Vancouver: UBC Press, 2008.

Anisef, Paul, and Michael Lanphier, eds. *The World in a City.* Toronto: University of Toronto Press, 2003.

Appadurai, Arjun. *The Future as a Cultural Fact.* London: Verso, 2013.

Appiah, Kwame. "The Case for Contamination." *New York Times Magazine,* 1 January 2006. http://www.nytimes.com/2006/01/01/magazine//01cosmopolitan.

Asian News. "Monterey Park Tries to Find Itself" Daily Dose 02/29/08. Asianweek.com

Asian Television Network. www.asiantelevision.com/.

Aubry, Larry. "Black–Korean American Relations: An Insider's View." In *Los Angeles – Struggles toward Multiethnic Community,* edited by Edward T. Chang and Russell C. Leong, 149–56. Seattle: University of Washington Press, 1993.

Bagga, Geetika. "From the Komagata Maru to Six Sikh MPs in Parliament." *Our Diverse Cities,* no. 4 (Fall 2007), 161–5.

Bahrdt, Hans."Public Activity and Private Activity as Basic Forms of City Association." In *Perspectives on the American Community,* edited by Roland Warren, 78–85. Chicago: Rand McNally & Co.,1966.

Bailey, Margo L. "Cultural Competency and the Practice of Public Administration." In *Diversity and Public Administration,* 2nd ed., edited by Mitchell F. Rice, 171–88. Armonk: M.E. Sharpe, 2010.

Banting, Keith, Thomas Courchene, and Leslie F. Seidle. "Introduction." In *Belonging? Diversity, Recognition and Shared Citizenship in Canada*, edited by K. Banting, T.J. Courchene and L.F. Seidle, 1–38. Montreal: Institute of Public Policy, 2007.

Barret, Giles, Trever Jones, and David McEvoy. "Ethnic Minority Business Theoretical Discourse in Britain and North America." *Urban Studies* 33, no. 4 (1996).

Barrett, Paul. *American Islam*. New York: Farrar, Straus and Giroux, 2006.

Beresky, Samuel. "A Movable Feast." *Planning*, February 2011, 32.

Berg, Bruce F. *New York City Politics*. New Brunswick, NJ: Rutgers University Press, 2007.

Berger, Joseph. *The World in a City*. New York: Ballantine Books, 2007.

Berger, Joseph. "Protests of a Plan for a Mosque? That Was Last Year," *New York Times*, 19 August 2011, http://www.nytimes.com/2011/08/19/nyregion/mosque-opens-quietly-on-staten-island.html.

Berry, Brian J.L. "General Features of Urban Commercial Structure." In *Internal Structure of the City*, edited by Larry S. Bourne, 361–7. New York: Oxford University Press,1971.

Biles, John, Meyer Burstein, and James Frideres, eds. *Immigration and Integration in Canada in the Twenty-first Century*. Montreal: School of Policy Studies, Queen's University, McGill-Queen's University Press, 2008.

Bishop, Bill. *The Big Sort: Why the Clustering of Like-Minded America Is Tearing Us Apart*. New York: Houghton Mifflin, 2008.

Bissoondath, Neil. *Selling Illusions*. Toronto: Penguin, 1994.

Bloemraad, Irene. "Diversity and Elected Officials in the City of Vancouver." In *Electing a Diverse Canada: The Representation of Minorities and Women*, edited by Caroline Andrew, John Biles, Myer Siemiatycki, and Erin Tolly, 46–69. Vancouver: UBC Press. 2008.

Bloomfield, Jude, and Franco Bianchini. *Planning for the Intercultural City*. Stroud, UK: Comedia, 2004.

Boal, Frederick. "Urban Ethnic Segregation and the Scenario Spectrum." In *Desegregating the City*, edited by David P. Varady, 62–78. Albany: State University of New York Press, 2005.

Bonancich, Edna, and John Modell. *The Economic Basis of Ethnic Solidarity: Small Business in the Japanese American Community*. Berkeley: University of California Press,1980.

Borjas, George J. "The Impact of Immigrants on Employment Opportunities of Natives." In *Immigration Reader*, edited by David Jacobson, 217–30. Malden: Blackwell Publishers, 1998.

Bouchard, Gerard, and Charles Taylor. *Building the Future*. Report of Commission de Consultation sur les Pratiques d'Accommodement Reliées aux Différences Culturelles. Quebec: Bibliothèque et Archives Nationales du Québec, 2008.

Bourne, Larry S. "Patterns: Descriptions of Structure and Growth." In *Internal Structure of the City*, edited by Larry S. Bourne, 69–74. New York: Oxford University Press.1971.

Brettell, Caroline B. "Incorporating New Immigrants in a Sunbelt Suburban Metropolis." In *Twenty-First Century Gateways*, edited by Susan W. Hardwick and Caroline B. Brettell,75–9. Washington: Brookings Institution Press, 2008.

Brooks, David. *Bobos in Paradise: The New Upper Class and How They Got There*. New York: Simon and Schuster, 2001.

Brooks, David. "The Death of Multiculturalism." *New York Times*, 27 April 2001, http://query.nytimes.com/gst/fullpage.html?res=9C01E1DC133FF93 4A15757C0A9609C8B63.

Brooks, David. "The Sidney Awards, Part 1." *New York Times*, 29 December 2011. http://www.nytimes.com/2013/12/27/opinion/brooks-the-sidney -awards.html.

Brooks, Michael P. *Planning Theory for Practitioners*. Chicago: Planners Press, 2002.

Brown, Ian. "Vancouver Suddenly Buttons Down Its Shirt and Irons Its Ties." *Globe and Mail*, 30 January 2010, F3.

Buchanan, Pat. "On Immigration." *On the Issues*, http://www.ontheissues. org/Celeb/Pat_Buchanan_Immigration.htm.

Burayidi, Michael. "Tracking the Planning Profession: From Monistic Planning to Holistic Planning for a Multicultural Society." In *Urban Planning in a Multicultural Society*, edited by Michael Burayidi, 37–52. Westport, CT: Praeger, 2000.

Burayidi, Michael. "Urban Planning as a Multicultural Canon." Iin *Urban Planning in a Multicultural Society*, ed. Burayidi, 3–14. Westport: Praeger, 2000.

Burgess, Ernest W. "The Growth of the City." In *The City*, edited by Robert E. Park, Ernest W. Burgess, and Robert D. McKenzie, chap. 2. Chicago: University of Chicago Press, 1925.

Burns, Peter. *Electoral Politics Is Not Enough*. Albany: State University of New York Press, 2006.

Butter, Frank A., Erno Masurel, and Robert H.J. Mosch. "The Economics of Co-ethnic Employment Incentives, Welfare Effects and Policy Options." In *Handbook of Research on Ethnic Minority Entrepreneurship*, edited by Leo-Paul Dana, 55–6. Cheltenham: Edward Elger, 2007.

Canada Mortgage and Housing Corporation (CMHC). "2006 Census Housing Series: Issue 14 – The Housing Conditions of Visible Minority Households." *Research Highlight, Socio-economic Series 11–009.* October 2011.

Canada Mortgage Housing Corporation (CMHC). "Research Highlights: Socio-economic Series." Series 08-016 and 10-016. October 2010.

Canadian Institute of Planners. "*Planning Is.*" http://www.cip-icu.ca/web/la/en/pa/3FC2AFA9F72245C4B8D2E709990D58C3/template.asp.

Canadian Multiculturalism Act. Assented to 21 July 1988.

Caribana Arts Group. http://www.caribana.com.

Castells, Manuel. *City, Class and Power.* London: Macmillan Press,1978.

Charles, Camille Z. *Won't You Be My Neighbor?* New York: Russell Sage Foundation, 2006.

Chee, Maria W.L., Gary A. Dymski, and Wei Li. "Asia in Los Angeles." In *Chinese Entreprise, Transnationalism and Identity,* edited by Edmund T. Gomez and Hsin-Huang M. Hsiao, 217–19. London: Routledge Curzon, 2004.

Cheng, Lucie, and Yang, Philip. "Asians: The 'Model Minority' Deconstructed." In *Ethnic Los Angeles,* edited by Roger Waldinger and Mehdi Bozorgmehr, 305–44. New York: Russell Sage, 1996.

Chui, Tina, Kelly Tran, and John Flanders. "Chinese Canadians Enriching the Cultural Mosaic." *Statistics Canada Catalogue Number 11–006.* Spring 2005.

City of New York. "Total Population by Mutually Exclusive Race and Hispanic Origins." Table PL-P2A NYC. http://www.nyc.gov/html/dcp/html/census/dem-_tables_201.shtml.

City of Toronto. "Multilingual Services Policy." http://www.toronto.ca/legdocs/2002/agendas/council/cc020213/adm2rpt/cl004.pdf.

City of Toronto. "Street Food Pilot Project Update and Recommendations." Staff report, 2011. http://www.toronto.ca/legdocs/mmis/2011/ex/bgrd/backgroundfile-37387.pdf.

Clark, Terry Nichols. "The New Chicago School." In *The City Revisited,* edited by Dennis R. Judd and Dick Simpson, 220–41. Minneapolis: University of Minnesota Press, 2011.

Collacot, Martin. "Submission to the Commission on Accommodation Practices Related to Cultural Differences." Vancouver, Fraser Institute, 2007. http://www.Fraserinstitute.org/research_news/display.aspx/id=13433.

Crains New York Business. Press release. "Immigrants Playing a Larger Role in City's Economy." 13 January 2010.

Daily Dose. "Monterey Park Tries to Find Itself. " http://www.asianweek.com/2008/02/29/5062/.

Dana, Leo-Paul, editor. *Handbook of Research on Ethnic Minority Entrepreneurship.* Cheltenham: Edward Elger, 2007.

Database of Masjids, Mosques, and Islamic Centers. http://www.religioninsights
.org/resources/database-masjids-mosques-and-islamic-centers-us.

Davidoff, Paul. "Advocacy and Pluralism in Planning." *Journal of the American Institute of Planners* 31, no. 4 (November 1965), 331–8.

Davis, Mike. *City of Quartz*. London: Verso,1990.

Davis, Mike. *Magical Urbanism*. London: Verso, 2000.

Dear, Michael, and Dahmann, Nicholas. "Urban Politics and the Los Angeles School of Urbanism." In *The City Revisited*, edited by Dennis R. Judd and Dick Simpson, 65–78. Minneapolis: University of Minnesota Press, 2011.

Doob, Christopher B. *Race, Ethnicity and the American Urban Mainstream*. Boston: Pearson, 2005.

Drier, Peter, John Mollenkopf, and Todd Swanstorm. *Place Matters*. Lawrence: University of Kansas Press, 2004.

Duncan, Howard. "Multiculturalism: Still a Viable Concept for Integration." *Canadian Diversity* 4, no. 1 (Winter 2005).

Durkheim, Emile. *The Division of Labor in Society* (1893). Trans. George Simpson. New York: The Free Press, 1933.

Edwards, Mary E. *Regional and Urban Economics and Economic Development*. Boca Raton, FL: Auerbach Publications, 2007.

Espenshade, Thomas J. "U.S. Immigration and the New Welfare State." In *Immigration Reader*, edited by David Jacobson, 231–50. Malden: Blackwell Publishers, 1998.

Fainstein, Susan. *The Just City*. Ithaca: Cornell University Press, 2010.

Fessenden, Ford, and Sam Roberts. "Then as Now – New York's Shifting Ethnic Mosaic." *New York Times* (Region), 22 January 2011.

Fiscal Policy Institute. *Immigrants and the Economy*. New York: Fiscal Policy Institute, 2009.

Fisher, Claude. "The Subcultural Theory of Urbanism: A Twentieth-Year Assessment." *American Journal of Sociology* 101, no. 3 (November 1995), 543–77.

Fleras, Angie. *The Politics of Multiculturalism*. New York: Palgrave, 2009.

Florida, Richard. *The Flight of the Creative Class*. New York: Collins, 2005.

Foner, Nancy. "How Exceptional Is New York? Migration and Multiculturalism in the Empire City." *Ethnic and Racial Studies* 30, no. 6 (November 2007), 999–1023.

Foner, Nancy. *In a New Land*. New York: New York University Press, 2005.

Foner, Nancy, ed. *One Out of Three*. New York: Columbia University Press, 2013.

Fong, Eric. "A Comparative Perspective on Racial Residential Segregation: American and Canadian Experiences." *Sociological Quarterly* 37, no. 2 (1996), 199–226.

Fong, Eric, and Linda Lee. "Chinese Ethnic Economy within the City Context." In *Chinese Ethnic Business, Global and Local Perspectives*, edited by Eric Fong and Chiu Luk, 149–72. London: Routledge, 2007.

Fong, Eric, and Emi Ooka. "The Social Consequences of Participating in the Ethnic Economy." *International Migration Review* 36 (2002).

Fong, Eric, and Kumiko Shibuya. "Multiethnic Cities in North America." *Annual Review of Sociology* 31 (2005), 285–304.

Fong, Eric, and Rima Wilkes. "The Spatial Assimilation Model Reexamined: An Assessment of Canadian Data." *International Migration Review* 33, no. 3 (1999), 594–620.

Foot, David K. *Boom, Bust and Echo*. Toronto: Macfarlane Walter & Ross, 1996.

Foster, John. "Multicultural Planning in Deed: Lessons from the Mediation Practice of Shirley Solomon and Larry Sherman." In *Urban Planning in a Multicultural Society*, edited by Michael Burayidi, 147–68. Westport, CT: Praeger, 2000.

Frey, William H. *"Melting Pot Cities and Suburbs: Racial and Ethnic Change in Metropolitan America in the 2000s."* Metropolitan Policy Program at Brookings. http://www.brookings.edu?papers/2011/05/04_census_ethnic _frey.Aspx.1.

Frey, William H. "Diversity Explosion: The Cultural Generation Gap Mapped," 19 November 2014. www.brookings.edu.

Frey, William H. "New Projections Point to a Majority Minority Nation in 2044." In *The Avenue: Rethinking Metropolitan America*. 12 December 2014. http://www.brookings.edu/blogs/the-avenue/posts/2014/12/12 -majority-minority-nation-2044-frey.

Frideres, James. "Creating an Inclusive Society: Promoting Social Integration in Canada." In *Immigration and Integration in Canada in the Twenty-First Century*, edited by John Biles, Meyer Burstein, and James Frideres, 77–102. Montreal: McGill-Queen's University Press, 2008.

Frienson, Joe. "The GTA's Matchless Sports." *Globe and Mail*. 14 August 2011, M1.

Frisken, Frances, and Marcia Wallace. "Governing the Multicultural Region." *Canadian Public Administration* 46, no. 2 (2003), 153–77.

Frow, J., and M. Morris. *Australian Cultural Studies: A Reader*. St Leonards, NSW: Allen and Unwin, 1993.

Gans, Herbert. "The American Kaleidoscope, Then and Now." In *Reinventing the Melting Pot*, edited by Tamar Jacoby, 36–45. New York: Basic Books, 2004.

Gans, Herbert. *The Urban Villagers*. New York: Free Press, 1962.

Gladstone, David, and Susan S. Fainstein. "The New York and Los Angeles Economies from Boom to Crisis." In *New York and Los Angeles*, edited by David Halle and Andrew Beveridge.

Geertz, Clifford. *Peddlers and Princes: Social Development and Economic Change in Two Indonesian Towns*. Chicago: University of Chicago Press, 1963.

Giles, Barrett, Trever Jones, and David McEvoy. "Ethnic Minority Business: Theoretical Discourse in Britain and North America." *Urban Studies* 33, nos. 4–5 (1996), 788–809.

Glazer, Nathan. "Assimilation Today: Is One Identity Enough?" In *Reinventing the Melting Pot*, edited by Tamar Jacoby, 61–73. New York: Basic Books, 2004.

Glazer, Nathan. *We Are All Multiculturalist Now*. Cambridge, MA: Harvard University Press, 1997.

Glazer, Nathan, and Daniel Moynihan. *Beyond the Melting Pot*. Cambridge, MA: MIT Press, 1963.

Globe and Mail. "Loblaws Bought T and T." 25 February 2011, B5.

Globe and Mail. "Religion Topples Multiculturalism." 5 October 2010, A10.

Globe and Mail. "The Rise and Fall of the Ethnic Mall." 16 June 2012, M1 and M5.

Globe and Mail. "Vancouver Cheers Chinese Driven Real Estate Boom." 12 March 2011, A4.

Goering, John. "Introduction." *Fragile Rights within Cities*. Lanham, MD: Rowman & Littlefield, 2007.

Gonthier, Charles D. "Liberty, Equality and Fraternity: The Forgotten Leg of Trilogy or Fraternity the Unspoken Third Pillar of Democracy." *McGill Law Journal* 45 (2000).

Good, Kristen. "Multicultural Democracy in the City: Explaining Municipal Responsiveness to Immigrants and Ethno-Cultural Minorities." PhD thesis, University of Toronto, 2006.

Good, Kristin. "Patterns of Politics in Canada's Immigrant-receiving Cities and Suburbs." *Policy Studies* 26, no. 3/4 (2005), 261–88.

Graham, Katherine, and Susan Phillips. "Another Fine Balance: Managing Diversity in Canadian Cities." In *Belonging? Diversity, Recognition and Shared Citizenship in Canada*, edited by Keith Banting, Thomas Courchene, and F. Leslie Seidle, 155–95. Montreal: Institute for Research on Public Policy, 2007.

Grant, David M., Melvin L. Oliver, and Angela D. James. "African Americans: Social and Economic Bifurcation." In *Ethnic Los Angeles*, edited by Roger Waldinger and Mehdi Bozorgmehr, 379–412. New York: Russell Sage Foundation, 1996.

Green, James. *Cultural Awareness in Human Services*. Boston: Allyn and Bacon, 1999.

Grey, Julius. "The Paradoxes of Reasonable Accommodation." *Policy Options* 28, no.3 (2007), 34–6.

Hall, Peter. *Cities in Civilization*. London: Weidenfeld & Nicolson. 1998.

Halle, David, and Andrew A. Beveridge, eds. *New York and Los Angeles*. New York: Oxford University Press, 2013.

Halle, David, Eric Vanstrom, et al. "New York, Los Angeles and Chicago as Depicted in Hit Movies." In *New York and Los Angeles*, edited by David Halle and Andrew Beveridge.

Han, Chong-Suk. "We Both Eat Rice, but That Is About It: Korean and Latino Relations in Multi-Ethnic Los Angeles." In *Everyday Multiculturalism*, edited by Amanda Wise and Selvaraj Velayutham, 235–48. Houndsmill: Palgrave Macmillan, 2007.

Harlem World. "New $100 Million Development Planning for 125th Street in Harlem." 2000.

Harris, Amy Lavender. *Imagining Toronto*. Toronto: Mansfield Press, 2010.

Harvey, David. "The Right to the City." *International Journal of Urban and Regional Research* 27, no. 4 (December 2003).

Harwood, Stacy. "Struggling to Embrace Difference in Land-Use Decision-Making in Multicultural Communities." *Planning Practice and Research* 20, no. 4 (November 2005), 363–4.

Hawthorne, Christopher. "Latino Urbanism Influences or Los Angeles in Flux." *Los Angeles Times*, 6 December 2014.

Healy, Patsy. *Collaborative Planning*. Vancouver: UBC Press, 1997.

Henke, Holger, and George J.A. Irish. "Relations between the Jewish and Caribbean American Communities in New York City: Perceptions, Conflict and Co-operation." In *Race and Ethnicity in New York City*, edited by Jerome Krase and Ray Hutchison, 198–220. Amsterdam: Elsevier, 2004.

Hequet, Marc. "God and Mall." *Planning*, July 2010, 13–15.

Hiebert, Daniel. *Exploring Minority Enclave Areas in Montreal, Toronto, and Vancouver*. Citizenship and Immigration Canada, Research and Evaluation Branch, March 2009.

Hiebert, Daniel, Nadine Schuurman, and Heather Smith. *Multiculturalism "on the Ground": The Social Geography of Immigrant and Visible Minority Populations in Montreal, Toronto, and Vancouver, Projected to 2017*. Working paper no. 07–12. Metropolis British Columbia, Centre of Excellence for Research on Immigration and Diversity, April 2007.

Hochschild, Jennifer, and John Mollenkopf. "Modeling Immigrant Political Incorporation," in *Bringing Outsiders In*, edited by Jennifer L. Hochschild and John F. Mollenkopf, 4–24. Ithaca: Cornell University Press, 2009.

Hochschild, Jennifer, and John Mollenkopf. "Understanding Immigrant Political Incorporation through Comparison." In *Bringing Outsider In*, edited by J. Hochschild and J. Mollenkopf.

Hodge, Gerald, and David Gordon. *Planning Canadian Communities*. Scarborough, ON: ITP/Nelson, 2007.

Hollinger, David. *Postethnic America: Beyond Multiculturalism*. New York: Basic Books, 2000.

Holston, James, and Arjun Appadurai. "Cities and Citizenship." *Public Culture* 8 (1996).

Hou, Feng, and Garnett Picot. "Visible Minority Neighbourhoods in Toronto, Montreal and Vancouver." *Canadian Social Trends*, Spring 2004, 8–13.

Hoyt, Homer. *The Structure and Growth of Residential Neighborhoods in American Cities*. Chicago: University of Chicago Press, 1939.

Huber, Cheryl. "New York Wants to Engage Immigrants in Its Parks." New Yorkers for Parks, canada.metropolis.net/pdfs/fow_30aug10_huber_e.pdf.

Hulchanski, David. "The Three Cities within Toronto: Income Polarization among Toronto's Neighbourhoods, 1970–2000." Bulletin # 41. University of Toronto, Centre of Urban and Community Studies, 2007.

Hum, Terry. " Immigrant Global Neighborhood in New York City." In *Race and Ethnicity in New York City*, edited by Jerome Krase and Ray Hutchison, 25–38. Amsterdam: Elsevier, 2004.

Hum, Tarry. "Immigration Grows to Half of New York's Labour Force." *Regional Labour Review*, Spring 2005.

Huntington, Samuel. *Who Are We? America's Great Debate*. New York: The Free Press, 2005.

Hutter, Mark. *Experiencing Cities*. Boston: Pearson, 2007.

Innes, Judith. "Information in Communicative Planning." *Journal of the American Planning Association* 64, no. 1 (January 1998), 52–63.

Isaac, Reginald. "The Dynamics of Urban Renewal." In *Taming Metropolis*, edited by Wentworth Elderedge, 784–98. New York: Anchor Books, 1967.

Iver, Pico. *The Global Soul*. New York: Viking Books, 2000.

Jacobs, Jane. *Cities and the Wealth of Nations*. Harmondsworth: Viking, 1984.

Jacobs, Jane. *The Death and Life of Great American Cities*. New York: Vintage Books, 1992.

Jacobson, Matthew. *Whiteness of a Different Color: European Immigrants and the Alchemy of Race*. Cambridge, MA: Harvard University Press, 1998.

Kaisinitz, Philip, John H. Mollenkopf, Mary C. Waters, and Jennifer Holdaway. *Inheriting the City*. New York: Russell Sage Foundation, 2008.

Kallen, Horace. *Cultural Pluralism and the American Idea: An Essay in Social Philosophy*. Philadelphia: University of Pennsylvania Press, 1956.

Kallick, David D. "Immigrants and Economic Growth in New York City." In *One Out of Three*, edited by Nancy Foner.

Kaplan, David H., and Wei Li. "Introduction: The Places of Ethnic Economies." In *Landscapes of the Ethnic Economy*, edited by David H. Kaplan and Wei Li, 1–16. Lanham, MD: Rowman and Littlefield, 2006.

Khaldun, Ibn. *The Muqaddimah*. Translated by Franz Rosenthal. Princeton: Princeton University Press, Bollingen series, 1965.

Kivisto, Peter. "We Really Are All Multiculturalists Now." *Sociological Quarterly* 6, no. 3 (2012).

Kloosterman, Robert, Joanne Van Der Leun, and Jan Rath. "Mixed Embeddedness: (In)formal Economic Activities and Immigrant Businesses in the Netherlands." *International Journal of Urban and Regional Research* 23, no. 2 (2001).

Klosterman, Richard E.K. "Planning Theory Education: Thirty Years Review" *Journal of Planning Education and Research* 31, no. 3 (2011).

Kumar, Sandeep, and Bonica Leung. "Formation of an Ethnic Enclave: Process and Motivation." *Plan Canada* 45, no. 2 (Summer 2005), 43–5.

Kymlicka, Will. "Ethnocultural Diversity in a Liberal State: Making Sense of Canadian Model(s)." In *Belonging? Diversity, Recognition and Shared Citizenship in Canada*, edited by Keith Banting, Thomas J. Courchene, and F. Leslie Seidle, 39–86. Montreal: Institute of Public Policy, 2007.

Kylmicka, Will. *Finding Our Way*. Toronto: Oxford University Press, 1998.

Kymlicka, Will. *Multicultural Odysseys*. New York: Oxford University Press, 2007.

Lal, Vinay. "Sikh Kirpans in California Schools: The Social Construction of Symbols, Legal Pluralism, and the Politics of Diversity." *Amerasia Journal* 23, no. 1 (1996), 57–89.

Laslett, John H. "Historical Perspectives: Immigration and the Rise of a Distinctive Urban Region, 1900–1970." In *Ethnic Los Angeles*, edited by Roger Waldinger and Mehdi Bozorgmehr, 39–78. New York: Russell Sage Foundation, 1996.

Lee, Jennifer. *Civility in the City*. Cambridge, MA: Harvard University Press, 2002.

Lefebvre, Henri. *Writings on Cities*. Translated by Eleonore Kofman and Elizabeth Lebas. Malden: Blackwell Publishing, 2006.

Legrain, Philippe. *Immigrants: Your Country Needs Them*. London: Abacus, 2007.

Lewis, John. "A Poll Tax by Another Name." *New York Times*, 27 August 2011.

Ley, David F., and Larry S. Bourne. "Introduction: The Social Context and Diversity of Urban Canada." In *The Changing Social Geography of Canadian Cities*, edited by Larry S. Bourne and David F. Ley, 2–14. Montreal: McGill-Queen's University Press, 1993.

Ley, David. *Millionaire Migrants: Trans-Pacific Lifelines*. London: Wiley Blackwell, 2010.

Li, Peter. "Ethnic Enterprise in Transition: Chinese Businesses in Richmond B.C. 1980–90." *Canadian Ethnic Studies*, no. 24 (1992), 120–38.

Li, Wei, and Gary Dymski. "Globally Connected and Locally Embedded Financial Institutions." In *Chinese Ethnic Business: Global and Local Perspectives*, edited by Eric Fong and Chiu Luk, 35–59. London: Routledge, 2007.

Light, Ivan. "Globalization, Transnationalism and Chinese Transnationalism." In *Chinese Ethnic Business: Global and Local Perspectives*, edited by Eric Fong and Chiu Luk, 89–98. London: Routledge, 2007.

Light, Ivan, and Steven J. Gold. *Ethnic Economies*. San Diego: Academic Press, 2000.

Light, Ivan, and Elizabeth Roach. "Self-employment: Mobility Ladder or Economic Life Boat?" In *Ethnic Los Angeles*, edited by Roger Waldinger and Mehdi Bozorgmehr, 193–214. New York: Russell Sage Foundation, 19

Lin, Jan. *The Power of Urban Ethnic Places*. New York: Routledge, 2011.

Lin, Jan. *Reconstructing Chinatown*. Minneapolis: University of Minnesota Press, 1998.

Lo, Lucia, and Shuguang Wang. "The New Chinese Business Sector in Toronto: A Spatial and Structural Anatomy of Medium-Sized and Large Firms." In *Chinese Ethnic Business: Global and Local Perspectives*, edited by Eric Fong and Chiu Luk, 64–87. London: Routledge, 2007.

Lobo, Arun, and Joseph J. Salvo. "A Portrait of New York City." In *Race and Ethnicity in New York City*, edited by Ray Hutchison and Jerome Krase. Amsterdam: Elsevier, 2007.

Logan, John, and John Mollenkopf. *People Politics in American Big Cities*. New York: Drum Major Institute for Public Policy, 2003.

Logan, John R., Wenquen Zhang, and Richard D. Alba. "Immigrant Enclaves and Ethnic Communities in New York and Los Angeles." *American Sociological Review* 67, no. 2 (April 2002), 299–322.

Logan, John, and Zhang Wei Wei. *Separate but Equal: Asian Nationalities in the US*. Brown University, 2010. http://www.s4.brown.edu/us2010/data/Report/report06112013.pdf.

Lorinc, John. *The New City*. Toronto: Penguin Canada, 2008.

Lorinc, John. "The New Regent Park," *UofTMagazine*, Spring 2013, http://magazine.utoronto.ca/feature/new-regent-park-toronto-community-housing-john-lorinc/.

Los Angeles Economic Development Commission. *Los Angeles Statistics 2008*.

Maclean's. "Too Asian." 22 November 2010, 15.

Madood, Tariq. "Multiculturalism. Ethnicity and Integration: Contemporary Challenges." *Canadian Diversity* 5, no. 1 (Winter 2006), 120–2.

Madood, Tariq. *Multiculturalism*. Princeton: Princeton University Press, 2007.

Marcuse, Peter. "Enclaves Yes, Ghetto No: Segregation and the State." In *Desegregating the City*, edited by David P. Varady, 15–30. Albany: State University of New York Press, 2005.

Marotta, Vince. "Multicultural and Multiethnic Cities in Australia." In *Ethnic Landscape in an Urban World*, edited by Ray Hutchison and Jerome Krase, 38–46. Amsterdam: Elsevier, 2007.

Martin, Lawrence."Enough of Multiculturalism – Bring on the Melting Pot." *Globe and Mail*, 31 March 2009, A17.

Massey, Douglas S., and Nancy A. Denton. *American Apartheid: Segregation and the Making of the Underclass*. Cambridge, MA: Harvard University Press, 1993.

Massey, Douglas, and Nancy Denton. "Spatial Assimilation as a Socioeconomic Outcome." *American Sociological Review* 50, no. 1 (1985), 94–106.

Matthews, Christine M. "Foreign Science and Engineering Presence in U.S. Institutions and the Labor Force." Congressional Research Service Report 7-5700, 4–5.

Maytree Foundation. *Diversity on Board.* http://apps.maytree.com/diversecity/.

McCarron, John. "Majoring in Development." *Planning*. October 2011, 13.

Metropolitan Program at Brookings. *The State of Metropolitan America, 2010.*

Mills, C. Wright. *The Power Elite*. London: Oxford University Press, 1956.

Milroy, Beth Moore, and Marcia Wallace. *Ethnoracial Diversity and Planning Practices in the Greater Toronto Area*. Centre of Excellence for Research in Immigration and Settlement, University of Toronto, 2002.

Missionary Outlook (Toronto) 30, no. 12, 12 December 1910.

Mollenkopf, John Hull. "School Is Out." In *The City Revisited*, edited by Dennis R. Judd and Dick Simpson, 169–85. Minneapolis: University of Minnesota Press, 2011.

Mollenkopf, John, and Raphael Sonenshein. "The New Urban Politics of Integration: A View from the Gateway Cities." In *Bringing Outsiders In*, edited by Jennifer Hochschild and John Mollenkopf, 74–93. Ithaca: Cornell University Press, 2009.

Muller, Thomas. *Immigrants and the American City*. New York: New York University Press, 1993.

Mumford, Lewis. *The City in History: Its Origins, Its Transformations, and Its Prospects*. New York: Harcourt, Brace and Co., 1961.

Murdie, Robert. "Diversity and Concentration in Canadian Immigration: Trends in Toronto, Montreal and Vancouver 1971–2006." Research Bulletin #42. University of Toronto, Centre of Urban and Community Studies, 2008.

Murdie, Robert, and Carlos Teixeira. "Towards a Comfortable and Appropriate Housing: Immigrant Experiences in Toronto." In *The World in a City*, edited by Paul Anisef and Michael Lanphier, 132–91. Toronto: University of Toronto Press, 2003.

Myers, Dowell. *Immigrants and Boomers*. New York: Russell Sage Foundation, 2007.

Myles, John, and Hou Feng. "Changing Colours: Spatial Assimilation and New Racial Minority Immigrants." *Canadian Journal of Sociology* 29, no. 1 (March 2004), 29–58.

Napoli, Philip. "Little Italy: Resisting the Asian Invasion 1965–96" In *Race and Ethnicity in New York City*, edited by Jerome Krase and Ray Hutchison, 246–59. Amsterdam: Elsevier, 2004.

New York City, Department of City Planning. *The Newest New Yorkers 2013*. New York Department of City Planning, Population Division, 2013.

New York City, Department of City Planning. Table PL-P2A NYC. http://www.nyc.gov/html/dcp/html/census/demo_tables_2010shtml.

New York Times. "India Hitching Hopes on Subway." 14 May 2010. http://www.nytimes.com/glogin?URI=http%3A%2F%2Fwww.nytimes.com%2F2010%2F05%2F14%2Fworld%2Fasia%2F14delhi.html%3F_r%3.

New York Times. "In Neighborhood That's Diverse, a Push for Signs to Be Less." 2 August 2011, www.nytimes.com?2011/08/02/nyregion/queens-councilman-wants-english-signs. html.

New York Times. "Little Known Guide Helps Police Navigate a Diverse City." 10 June 2013. http://www.nytimes.com/2013/06/11/nyregion/a-not-for-tourists-guide-to-navigating-a-multicultural-city.htm.

New York Times. "Muslim Community Centre Park 51." [2011]. http://topics.nytimes.com/top/reference/timestopics/organizations/p/park51/index.html.

New York Times. "Playing a Sport with Bats and Balls, but No Pitcher." 3 April 2008. http://www.nytimes.com/glogin?URI=http://www.nytimes.com/2008/04/03/nyregion/03cricket.html&OQ=_rQ3D0&OP=690e9591Q2FfkQ3CcfmucfSSSflcb-fQ5EkHtQ5DkkcgfgQ22Q22WfQ221fQ22VfmuQ5DrD.kmfQ22VHQ5D.HLrcQ25lcb

Ontario Municipal Board. *Tilzen Holdings Ltd versus Town of Markham*. 1998. O.M.B.D. no. 319, File no. PL 956623.D950049, Z960134, M970061.

Ornstein, Michael. *Ethno-Racial Groups in Toronto 1971–2001: A Demographic and Social-Economic Profile*. Toronto: Institute for Social Research, York University, 2006.

Ortiz, Vilma. "The Mexican-origin Population: Permanent Working Class or Emerging Middle Class?" In *Ethnic Los Angeles*, edited by Roger Waldinger and Mehdi Bozorgmehr, 247–78. New York: Russell Sage Foundation, 1996.

Pager, Devah, and Shepherd, Hana. "The Sociology of Discrimination: Racial Discrimination in Employment, Housing, Credit, and Consumer Markets." *American Review of Sociology* 34 (January 2008), 181–209.

Parekh Report, The. *The Future of Multiethnic Britain.* London: Profile Books 2000.

Peach, Ceri. "The Ghetto and the Ethnic Enclave." In *Desegregating the City,* edited by David Varady, 31–48. Albany: State University of New York Press, 2005.

Peart, Nicholas. "Why Is N.Y.P.D. after Me?" *New York Times,* 18 December 2011. www.nytimes/2011/12/18/opinion/sunday/young-black-and -frisked-by-the-nypd.html/ref=c.

Pessar, Patricia, and Pamela Graham. "Dominicans: Transnational Identities and Local Politics." In *New Immigrants in New York,* edited by Nancy Foner, 262–8. New York: Columbia University Press, 2001.

Pestieau, Katherine, and Marcia Wallace. "Challenge and Opportunities for Planning in the Ethno-Culturally Diverse City: A Collection of Papers – Introduction." In *Planning Practice and Theory* 4, no. 3 (September 2003), 253–8.

Pickett, Stanley. "An Appraisal of the Urban Renewal Programme in Canada." *University of Toronto Law Journal* 18, no. 3 (Summer 1968), 233–45.

Pitts, Sharifa. *Harlem Is Nowhere.* New York: Little, Brown and Company, 2011.

Poirier, Cecile, Annick Germain, and Amelie Billette. "Diversity in Sports and Recreation: A Challenge or an Asset for Municipalities of Greater Montreal?" *Canadian Journal of Urban Research,* special issue, vol. 15, no. 2 (2006, supplement), 38–49.

Polese, Mario, and Richard Stren. "Introduction." In *The Social Sustainability of Cities: Diversity and the Management of Change,* edited by Mario Polese and Richard Stren, 3–14. Toronto: University of Toronto Press, 2000.

Porter, John. *The Vertical Mosaic: An Analysis of Class and Power in Canada.* Toronto: University of Toronto Press, 1965.

Portes, Alejandro. *The New Second Generation.* New York: Russell Sage Foundation, 1996.

Portes, Alejandro, and Robert L. Bach. *Latin Journey: Cuban and Mexican Immigrants in the United States.* Berkeley: University of California Press, 1985.

Portes, Alijendro, and Ruben G. Rumbautm. *A Portrait of Immigrant America.* 3rd edition. Berkeley: University of California Press, 2006.

Poulson, M., J. Forrest, and R. Johnston. "From Modern to Post-Modern? Contemporary Ethnic Residential Segregation in Four U.S. Metropolitan Areas." *Cities* 19, no. 3 (2002), 161–72.

Purcell, Mark. "Possible Worlds: Henri Lefebvre and the Right to the City." *Journal of Urban Affairs* 36, no. 1 (2013).

Putnam, Robert. *Bowling Alone.* New York: Simon and Schuster, 2000.

Qadeer, Mohammad A. "The Charter and Multiculturalism." *Policy Options* 28, no. 2 (February 2007).

Qadeer, Mohammad A. "Pluralistic Planning for Multicultural Cities: The Canadian Practice." *Journal of the American Planning Association* 63, no. 4 (Autumn 1997), 481–94.

Qadeer, Mohammad A. *Ethnic Malls and Plazas: Chinese Commercial Developments in Scarborough, Ontario.* CERIS Working paper. Toronto: Joint Centre of Excellence for Research on Immigration and Settlement, 1998.

Qadeer, Mohammad A. *The Bases of Chinese and South Asian Merchants' Entrepreneurship and Ethnic Enclaves.* CERIS Working paper no. 9. Toronto: Joint Centre of Excellence for Research on Immigration and Settlement, 1999

Qadeer, Mohammad A., and Maghfoor Chaudhry. "The Planning System and the Development of Mosques in the Greater Toronto Area." *Plan Canada* 40, no. 2 (February/March 2000), 17–21.

Qadeer, Mohammad A. "Urban Planning and Multiculturalism, Beyond Sensitivity." *Plan Canada* 40, no. 4 (September 2000), 16–18.

Qadeer, Mohammad A., and Sandeep Kumar. "Ethnic Enclaves and Social Cohesion." *Canadian Journal of Urban Research*, special issue (Our Diverse Cities: Challenges and Opportunities), vol. 15, no. 2 (2006), 1–17.

Qadeer, Mohammad. "What Is This Thing Called Multicultural Planning?" *Plan Canada* 28, no. 3 (2007).

Qadeer, Mohammad A., Sandeep K. Agrawal, and Alexander Lovell. "Evolution of Ethnic Enclaves in the Toronto Metropolitan Area, 2001–2006." *Journal of International Migration and Integration* 11 (2010), 315–39.

Qadeer, Mohammad A., and Sandeep K. Agrawal. "The Practice of Multicultural Planning in American and Canadian Cities." *Canadian Journal of Urban Research* 20, no. 1 (supplement, 2011), 132–56

Rath, Jan. "Debating Multiculturalism: Europe's Reaction in Context." *Harvard International Review*, 6 January 2011. http://hir.harvard.edu.

Reeves, Dory. *Planning for Diversity.* London: Routledge, 2005.

Reitz, Jeffery G. *The Survival of Ethnic Groups.* Toronto: McGraw-Hill Ryerson, 1980.

Reitz, Jeffery G. "Ethnic Concentrations in Labour Markets and Their Implications for Ethnic Inequality." In *Ethnic Identity and Equality*, edited by Raymond Breton, Wsevolod W. Isajiw, et al. Toronto: University of Toronto Press, 1990.

Reitz, Jeffery G., and Rupa Banerjee. "Racial Inequality, Social Cohesion and Policy Issues." In *Belonging? Diversity, Recognition and Shared Citizenship in*

Canada, edited by Keith Banting, Thomas J. Courchene, and F. Leslie Seidle, 489–546. Montreal: Institute of Research in Public Policy, 2007.

Rex, John. "The Concept of a Multicultural Society." In *The Ethnicity Reader*, edited by Montserret Guibernam and John Rex, 205–28. Cambridge: Polity Press, 1997.

Rex, John. *Ethnic Minorities in the Modern Nation State*. Houndmills: MacMillan Press, 1996.

Ricci et al. v. DeStefano et al. http://www.supremecourt.gov/opinions/08pdf/07-1428.pdf.

Rice, Mitchell E. "Cultural Competency, Public Administration and Public Service Delivery in an Era of Diversity." In *Diversity and Public Administration*, edited by Mitchell F. Rice, 189–209. Armonk, NY: M.E. Sharpe, 2010.

Roseman, Curtis C., Hans D. Laux, and Gunter Thieme. "Preface." In *EthniCity*, edited by Curtis C. Roseman, Hans D. Laux, et al., v–xviii. London: Rowman and Littlefield Publishers, 1996.

Rosenbaum, Emily, and Samantha Friedman. *The Housing Divide*. New York: New York University Press, 2007.

Sanchez-Jankowski, Martin. *Cracks in the Pavement*. Berkeley: University of California Press, 2008.

Sandercock, Leonie. *Cosmopolis II Mongrel Cities*. London: Continuum, 2003.

Sandercock, Leonie. "Integrating Immigrants: The Challenge for Cities, City Governments and the City-Building Professions." In *Intercultural City*, edited by Phil Wood, 258–72. Stroud: Comedia, 2004.

Sandercock, Leonie. "Towards a Planning Imagination for the 21st Century." *Journal of the American Planning Association* 70, no. 2 (Spring 2004), 133–41.

Sandercock, Leonie, and Ann Forsyth. "A Gender Agenda: New Directions for Planning Theory." *Journal of the American Planning Association* 58, no. 1 (Winter 1992), 49–59.

Santos, Milton. *The Shared Space*. London: Methuen, 1979.

Sardar, Ziauddin. *Balti Britain*. London: Granta, 2008.

Sassen, Saskia. *The Global City*. Princeton: Princeton University Press, 1991.

Saunders, Doug. *Arrival City*. Toronto: Alfred A. Knopf, 2010.

Savage, Mike, Alan Warde, and Kevin Ward. *Urban Sociology, Capitalism and Modernity*. 2nd edition. Houndmills: Palgrave Macmillan, 2003.

Schrag, Peter. "California's Future Homeowners." *Los Angeles Times*, 27 July 2011. http://articles.latimes.com/2011/jul/27/opinion/la-oe-schrag-immigrants-20110727.

Scott, Allen. *Metropolis: From the Division of Labour to Urban Form*. Berkeley: University of California Press, 1988.

Sen, Amartya. *Identity and Violence.* New York: W.W. Norton, 2006.

Sennet, Richard. *The Conscience of the Eye.* London: W.W. Norton,1990.

Siemiatycki, Myer. *The Diversity Gap: The Electoral Under-representation of Visible Minorities.* Toronto: Ryerson Centre of Immigration and Settlement, 2011.

Siemiatycki, Myer. "Reputation and Representation: Reaching for Political Inclusion." In *Electing a Diverse Canada: The Representation of Minorities and Women,* edited by Caroline Andrew, John Biles, Myer Siemiatycki, and Erin Tolly, 23–45. Vancouver: UBC Press, 2008.

Siemiatycki, Myer, Tim Rees, and Kahn Rahi. "Integrating Community Diversity in Toronto: On Whose Terms?" In *The World in a City,* edited by Paul Anisef and Michael Lanphier, 373–456. Toronto: University of Toronto Press, 2003.

Siemiatycki, Myer, and Anver Saloojee. "Ethnoracial Political Representation in Toronto: Patterns and Problems." *Journal of International Migration and Integration* 3, no. 2 (Spring 2002), 241–73.

Simmel, George."The Metropolis and Mental Life" (1903). In *Perspectives on the American Community,* edited by Roland L. Warren, 18–24. Chicago: Rand McNally and Co.,1966.

Singer, Audrey. "An Introduction." In *Twenty-First Century Gateways,* edited by Susan W. Hardwick and Caroline B. Brettell, 3–16.Washington, DC: Brookings Institution Press, 2008.

Statistics Canada. *2006 Census, Immigration in Canada: A Portrait of the Foreign-born Population, 2006 Census: Findings.* http://www12.statcan.ca/census -recensement/2006/as-sa/97-557/index-eng.cfm.

Statistics Canada. "Daily," 9 March 2010.

Stein, Janice G. "Searching for Equality." In *Uneasy Partners,* edited by Janice G. Stein, David R. Cameron, et al., 1–22. Waterloo: Wilfrid Laurier University Press, 2007.

Stewart, Ella."Communications between African Americans and Korean Americans before and after Los Angeles Riots." In *Los Angeles: Struggles toward Multiethnic Community,* edited by Edward Chang and Russell Leong, 23–54. Seattle: University of Washington Press. 1993.

Stone, Clarence. "Urban Regimes and Capacity to Govern: A Political Economy Approach." *Journal of Urban Affairs* 15, no. 1 (1993),1–28.

Suttles, Gerald. *The Social Order of the Slum: Ethnicity and Segmentation.* Chicago: University of Chicago Press, 1968.

Tajbakhsh, Kian. *The Promise of the City.* Berkeley: University of California Press, 2001.

The Parekh Report – The Future of Multiethnic Britain. London: Profile Books, 2000.

Thomas, Huw. "Race Equality and Planning: A Changing Agenda." *Planning, Practice and Research* 23, no. 1 (February 2008), 1–17.

Thompson, Richard H. *Toronto's Chinatown.* New York: AMS Press, 1989.

Thompson, Wilber. *A Preface to Urban Economics.* Baltimore: Johns Hopkins University Press, 1965.

Tindal, C. Richard, and Susan N. Tindal. *Local Government in Canada.* Scarborough, ON: Nelson Thompson Learning, 2000.

Tong, Benson. *The Chinese Americans.* Boulder: University Press of Colorado, 2003.

Toronto Life. "Toronto Might Be Moving to Food Cart Sanity, if Province, City Hall and Restaurant Owners Let It Happen." 11 April 2011.

Toronto Star. "High-Rise Ghettos." Life section, 3 February 2001, M1–3.

Toronto Star. "Known to Police." Insight section, 10 March 2012, IN4.

Toronto Star. "Stockton Is Biggest City in U.S. to File for Bankruptcy." 28 June 2012, A32.

Tseng, Yen-Fen. "Chinese Ethnic Economy: San Gabriel Valley, Los Angeles County." *Journal of Urban Affairs* 16, no. 2 (1994), 169–89.

Tsui, Bonnie. *American Chinatown.* New York: Free Press, 2009.

UNESCO. *Universal Declaration of Cultural Diversity.* 21 February 2002.

Valpy, Michael. "Our Part-Time Home and Native Land." *Globe and Mail,* 28 June 2008, A13.

Varshney, Ashtosh. *Ethnic Conflict and Civic Life.* New Haven: Yale University Press, 2002.

Viswanathan, Leela. "Postcolonial Planning and Ethno-Racial Diversity in Toronto: Locating Equity in Contemporary Planning Context." *Canadian Journal of Urban Research* 18, no. 1 (Summer 2009), 162–82.

Vitiello, Domenic. "The Migrant Metropolis and American Planning." *Journal of the American Planning Association* 25, no. 2 (Spring 2009), 245–56.

Volery, Thierry. "Ethnic Entrepreneurship: A Theoretical Framework." In *Handbook of Research on Ethnic Minority Entrepreneurship,* edited by Leo-Paul Dana.

Wadhwa, Vivek. "How the Indians Conquered in Silicon Valley." 13 January 2012. http://www.inc.com/vivek-wadhwa/how-the-indians-succeeded-in-silicon-valley.html.

Waldinger, Roger, Howard Aldrich, and Robin Ward. "Opportunities, Group Characteristics and Strategies." In *Ethnic Entrepreneurs: Immigrant Businesses in Industrial Societies,* edited by R. Waldinger, H. Aldrich, and R. Ward, 13–48. London: Sage, 1990.

Waldinger, Roger, and Yen-Fen Tseng. "Divergent Diasporas: The Chinese Communities of New York and Los Angeles Compared." *Revue Européenne des Migrations Internationales* 8 (1992), 91–116.

Waldinger, Roger, and Yen-Fen Tseng. "The Making of a Multicultural Metropolis." In *Ethnic Los Angeles*, edited by Roger Waldinger and Mehdi Bozorgmehr, 3–38. New York: Russell Sage Foundation, 1996.

Walks, R. Alan, and Larry Bourne. "Ghettos in Canada's Cities? Racial Segregation, Ethnic Enclaves and Poverty Concentration in Canadian Urban Areas." *The Canadian Geographer* 50, no. 3 (2006), 273–97.

Wallace, Marcia, and Beth Moore Milroy. "Interesting Claims: Possibilities for Planning in Canada's Multicultural Cities." In *Gender, Planning and Human Rights*, edited by Tony Fenster, 55–73. London: Routledge, 1999.

Wang, Lu. "An Investigation of Chinese Immigrant Consumer Behavior in Toronto, Canada." *Journal of Retailing and Consumer Services* 11 (2004).

Wang, Shuguang, and Jason Zhong. *Delineating Ethnoburbs in Metropolitan Toronto*. CERIS Working Paper no. 100. Toronto: CERIS, 2013.

Ward, David. "The Emergence of Central Immigrant Ghettos in American Cities 1840–1920." *Annals of the Association of American Geographers* 68, no. 2 (1968).

Watson, C.W. *Multiculturalism*. Philadelphia: Open University Press, 2000.

Wayland, Sarah V. *Integration of Peel Immigrant Discussion Paper*. Brampton: Region of Peel, Human Services Department, 2010.

Weber, Max. *The City*. Translated and edited by Dan Martindale. New York: The Free Press, 1958.

Weber, Max. *The Protestant Ethic and the Spirit of Capitalism* (1903). Translated by Talcott Parsons. Mineola: Dover Publications, 2003.

Weil, Kathleen, Minister of Justice, Province of Quebec. "Bill 94. An Act to Establish Guidelines Governing Accommodation Requests within the Administration and Certain Institutions." Quebec Official Publisher, 2010,

Wellman, Barry, and Barry Leighton. "Networks, Neighborhoods, and Communities." *Urban Affairs Quarterly* 14, no. 3 (March 1979), 363–99.

Whyte, William. *Street Corner Society: Social Structure of an Italian Slum*. Chicago: University of Chicago Press, 1943.

Wilson, Phil. "Planned Temecula Valley Mosque Draws Opposition." *Los Angeles Times*, 18 July 2010. articles.latimes.com/2010/jul/18/local/la-me -mosque-20100718.

Wilson, William J. *The Truly Disadvantaged: The Inner City, the Underclass, and Public Policy*. Chicago: University of Chicago Press, 1987.

Winnick, Louis. *New People in Old Neighborhoods*. New York: Russell Sage, 1990.

Wirth, Louis. "Urbanism as a Way of Life." *American Journal of Sociology* 44, no. 1 (July 1938), 8–20. Reprinted in *Perspectives of American Community*, edited by Roland Warren, 44–53. Chicago: Rand McNally,1966.

Wise, Amanda, and Selvaraj Velayutham. "Introduction." In *Everyday Multiculturalism*, edited by Amanda Wise and Selvaraj Velayutham, 3–15. Houndsmill: Palgrave Macmillan, 2007.

Wolfe, Jeanne. "Our Common Past: An Introduction to Canadian Planning History." *Plan Canada*. Special anniversary issue, 1994, 12–34.

Wood, Phil, and Charles Landry, editors. *The Intercultural City: Planning for Diversity Advantage*. London: Earthscan, 2008.

Woodsworth, J.S. *Strangers within Our Gates. Or: Coming Canadians* (1909). Toronto: University of Toronto Press, 1972.

Zhou, Min. *Contemporary Chinese America*. Philadelphia: Temple University Press, 200

Zhou, Min, Margaret Chin, and Rebecca Kim. "The Transformation of Chinese American Communities in New York and Los Angeles." In *New York and Los Angeles*, edited by David Halle and Andrew Beveridge.

Zhou, Yu. "How Do Places Matter? A Comparative Study of Chinese Economies in Los Angeles and New York City." *Urban Geography* 19, no. 6 (1998), 531–53.

Zukin, Sharon. *The Culture of Cities*. Oxford: Blackwell, 1995.

Index

demographics, 46; labour rights,
145; malls, 75; media, 159; political
representation, 178, 187; self-
employment rates, 102, 103, 105,
111. *See also* Muslims
parades, 81–2, 168, 191, 250
Parekh, B., 22, 24, 25, 29, 274
Parekh Report, 22
parking restrictions, 144, 213, 227,
229, 231
Peach, C., 66, 131
Philippines, 54
Phoenix (Arizona), 72
place, 37, 128–9, 172, 289n9
places of worship: in commercial
areas, 226–7, 303n36; as
community centres, 80; opposition
to, 72, 76; prayer rooms, 166, 208;
and social geography of ethno-
racial groups, 59, 76–81, *77, 80*, 87;
and urban planning, 78, 79, 218,
224–7, 234, 242, 302n34, 302–3n35,
303n36. *See also* gurudawaras;
mandirs; mosques
plazas, 74
pluralism, 25, 51, 214–15, 243, 246, 265
Poles, 40, 60, 130, 251
Polese, M., 86, 132
police services: and employment,
252; "Policing a Multicultural
Society" manual, 208, 252;
profiling, 40, 154, 164, 169, 172,
199, 254; and racism, 251
policy index of multicultural
planning, 223, 236–8
policymaking. *See* politics and
policymaking
political incorporation, 174–6, 190–4;
agency factors, 175–6; definition,
175, 295n9; minorities' strategies,

178, 181, 184, 189–90, 257;
non-electoral paths to, 188–90,
297nn42–3; structural factors,
175, 182
political representation: benefits of,
185–8, 252, 297n34; in Canada, 182,
185, 193; levels of, 192; models,
186, 297n34; and urban planning,
223; in US, 182, 193, 297n42. *See
also specific cities; specific ethnicities*
politics and policymaking, 171–94;
Canada, 184, 193, 296n22;
coalitions, 176, 181, 188; and
ethno-racial demands, 171, 172–4,
177, 187, 191, 193, 294–5n5, 296n22;
in immigrants' homelands, 178;
political incorporation, 174–6,
188–90, 190–4, 295n9, 297nn42–3;
representation benefits, 185–8,
297n34; United States, 193. *See also
specific cities*
Porter, J., 50
Portes, A., 59, 109
Portuguese: assimilation, 83;
enclaves, 46, 63, 64, 65, 66; ethnic
categorization, 259; statistics, 136*t*
post-modern city, 84, 85
poverty, and segregation, 60, 129,
133, 134. *See also* ghettos
power struggles, 154, 164–6, 170
prayer rooms, 166, 208, 209
preferential treatment, 212–13
pricing norms, 90
private domain, 21–3, 25, 33, 35, 128,
182
private-public domain relationships,
22–3
privatization of services, 121, 199, 211
profiling, 40, 154, 164, 169, 172,
199, 254

political representation, 180; self-employment rates, 102, 105, 111; transnational economies, 93, 116, 189

Tajbakhsh, K., 274

Tamils, 46, 259

Tammany Hall, 176

Target stores, 162

Taric mosque (Toronto), 79

taxi services, 74, 91, 94, 97, 103, 106, 122, 145, 249

Taylor, C., 24, 209, 237

technology industries, 89, 97, 98

telemarketing, 110

Temecula City (California), 225

temporary workers, 42, 54

terrorism, 20, 52, 78–9, 97, 265

Thompson, R., 114

Thrift, N., 29

Tim Hortons, 162

tokenism, 185

tolerance, 28, 52, 151, 154, 169, 293n23

Toronto: about, 9, 62–3, 87, 97, 98t, 143, 252; Chinese economies, 113–15, 117, 118–19, 287n54; consumer markets, 110; and diversity, 81, 191; earnings by ethnicity, 108; ethnic economic niches, 91, 97, 101–2, 105–6, 123; ethnic economies, 74; ethnic enclaves, 60–1, 62–7, 65, 72, 115, 132, 134, 142; ethnic groups, 10, 44, 45t, 46, 63; ethnic malls, 75; governance structure, 182, 183; housing, 66; iconic symbols, 260–1; immigrant demographics, 10, 44–6, 45t; integrated neighbourhoods, 73, 139, 140; as majority-minority city, 44–6, 45t, 113; as multicultural city, 5, 11, 252–4; opportunity structures, 97, 98t, 99, 100, 115, 118–19; places of worship, 79; polarization in, 106;

political representation in, 181–5, 192, 193, 206, 296n22, 296nn24–5, 296–7n29; social organization, 129, 147, 148–9; and social sustainability, 86; statistics, 9, 10, 44, 45t, 46, 97, 98t, 183–4, 200; and transnationalism, 93. *See also* entrepreneurship; Greater Toronto Area (GTA); Toronto Census Metropolitan Area (CMA); urban planning

Toronto Census Metropolitan Area (CMA): definition, 9, 61, 252; Dissimilarity Index, 136t; ethnic malls, 75; statistics, 9, 10, 62–3, 97, 98t, 102. *See also* Toronto

Toronto District School Board (TDSB), 210

tourism: and Chinese economies, 258; and civic culture, 168; as ethnic economic niche, 100; and ethnic enclaves, 59, 62; and ethnic malls, 59; opportunity structures, 97, 117, 123; and places of worship, 79

translation and interpretation services: and education, 209, 251; Language Access Services (New York), 204, 299n11; Language and Culture Resource Center (Los Angeles), 204; Language Line Service (Toronto), 203–4; Multilingual Access Program (Toronto), 203; multilingual access to services, 203–5, 214, 251, 257, 299n11, 299n13; Multilingual Service Unit (Toronto), 203–4; and urban planning, 223

transnationalism: and effects of information technology, 121; and ethnic economies, 92, 93, 105, 189;